AF702431

Entrenchment and Widening of the Upper San Pedro River, Arizona

Richard Hereford
U.S. Geological Survey
2255 North Gemini Drive
Flagstaff, Arizona 86001

THE GEOLOGICAL SOCIETY OF AMERICA · 1888 ·

SPECIAL PAPER

282

1993

Published by The Geological Society of America, Inc.
3300 Penrose Place, P.O. Box 9140, Boulder, Colorado 80301

Printed in U.S.A.

GSA Books Science Editor Richard A. Hoppin

Library of Congress Cataloging-in-Publication Data
Hereford, Richard.
Entrenchment and widening of the upper San Pedro River, Arizona / Richard Hereford.
p. cm. — (Special paper ; 282)
Includes bibliographical references.
ISBN 0-8137-2282-9
1. River channels—San Pedro River (Mexico and Ariz.) 2. River channels—Arizona. I. Title. II. Series: Special papers (Geological Society of America) ; 282.
GB1225.A6H47 1993
551.48'3'09791—dc20 93-6776
CIP

Cover photo: Inner valley of the San Pedro River, view upstream. Photograph from U.S. Bureau of Land Management.

10 9 8 7 6 5 4 3 2 1

Contents

Geological Society of America
Special Paper 282
1993

Entrenchment and Widening of the Upper San Pedro River, Arizona

Richard Hereford
U.S. Geological Survey, 2255 North Gemini Drive, Flagstaff, Arizona 86001

ABSTRACT

The San Pedro River of southeast Arizona is a north-flowing tributary of the Gila River. The area of the drainage basin upstream of the 40-km-long study reach is about 3,200 km^2. This study traces the historical evolution of the San Pedro River channel—specifically, the deepening, widening, and sediment deposition that have occurred since 1900—and it aims to evaluate the causes of channel widening and deepening, the rate of widening, and the present stability of the channel.

Alluvium of the river valley consists of upper Holocene pre- and postentrenchment deposits. The pre-entrenchment alluvium, which forms the principal terrace of the inner valley, accumulated between about A.D. 1450 and 1900 in a relatively sluggish, low-energy fluvial system with extensive marshy reaches and high water table. In contrast, postentrenchment alluvium, which forms the terrace, floodplain, and channel of the San Pedro River, was deposited in a relatively high-energy, entrenched, and meandering fluvial system.

The river flowed in a shallow, narrow channel on the surface of the unentrenched valley before 1890. A series of large floods, perhaps beginning as early as 1881, eventually led to entrenchment of the channel between 1890 and 1908. This deepening placed the channel 1 to 10 m below the former floodplain. The channel has widened substantially since entrenchment through lateral migration and expansion of entrenched meanders; its present size is 5.7 times greater than before entrenchment. The rate of channel expansion, however, has decreased since about 1955, coincident with a decrease of peak-flood discharge. Channel area increased at 0.1 km^2 yr^{-1} from entrenchment until 1955; since then the area increased at only 0.02 km^2 yr^{-1}, suggesting that the channel has stabilized and that further widening will probably be minor under present conditions of land use, discharge, and climate.

The reduction of peak-flow rates was related partly to increased channel sinuosity and to development of floodplains and riparian woodlands. The increased sinuosity produced a reservoir effect that attenuated flood waves, and the development of floodplains enabled flood waters to spread laterally, thereby increasing transmission losses. In addition, flow rates were probably affected by improved land use and changes of rainfall intensity and short-term rainfall patterns, which reduced runoff and decreased the time necessary for channel stabilization. Livestock grazing decreased steadily after the turn of the century, and numerous stock ponds and small water-retention structures were constructed in tributaries. The cumulative effect of these structures probably reduced peak-flow rates. Short-term rainfall patterns of the wet season (June 15–October 15) have probably changed from annual alternation of above- and below-average rainfall to a biennial or longer pattern. Moreover, frequency of low-intensity rainfall (daily rainfall

Hereford, R., 1993, Entrenchment and Widening of the Upper San Pedro River, Arizona: Boulder, Colorado, Geological Society of America Special Paper 282.

less than about 1.27 cm) was consistently above average for the decade 1957–1967. These factors probably improved conditions for growth and establishment of vegetation both in and outside of the channel.

The causes of the large floods that resulted in entrenchment are poorly understood, although climate and land use were key factors. Floods followed closely the rapid settlement of the area brought about by mining activity in the late 1870s; population rose from a few hundred to 6,000 in less than 5 yr. Extensive wood cutting for mine timber and fuel, suppression of wildfire, and reintroduction of large cattle herds undoubtedly exacerbated entrenchment. Flood-producing wet-season rainfall in the Southwest, however, was unusually heavy before, during, and shortly after entrenchment.

INTRODUCTION

The study area is the southern San Pedro River valley in Cochise County, southeast Arizona (Fig. 1). The San Pedro River is a north-flowing tributary of the Gila River, which drains 3,200 km^2 (upstream of the Charleston gage, Fig. 1) of the Basin and Range province in parts of northern Sonora, Mexico, and southeast Arizona. The studied 40-km-long reach includes most of the San Pedro Riparian National Conservation Area, which historically consisted of two Mexican land grants dating from the 1820s (Mattison, 1946) whose names are still preserved on modern maps (Fig. 1). The U.S. Bureau of Land Management administers the area for the protection and management of valuable riparian ecosystems, wildlife, and prehistoric and historic resources (Jackson and others, 1987).

Like most streams in the southwest United States, the San Pedro River has deepened and widened dramatically since the late 1800s. Once a narrow, unentrenched stream with extensive marshes (referred to as *cienegas*), beaver ponds, and abounding in fish, the river flows now 1 to 10 m below its former floodplain in a broad channel lined with cottonwood, willow, saltcedar, and mesquite. Fish and beaver and the marshes are gone, destroyed by the changing conditions of the entrenched channel and lowered water table. Despite this degraded condition, the inner valley of the San Pedro River is probably one of the richest wildlife habitats in the Southwest (Hunt, 1988). It is a nesting, migratory, or wintering habitat for 377 bird species and 35 raptor species, as well as an essential habitat for many other wildlife species, including 82 mammals (B. Lomeli, written communication, 1991). An extensive riparian woodland enhances the beauty of the area and provides the lush wildlife habitat. Ironically, the woodlands have developed since channel entrenchment, and their presence and expansion are closely linked to the widening process.

The geomorphic evolution of the San Pedro River channel since entrenchment, or since about 1900, was analyzed. The specific objectives were to evaluate the causes of entrenchment, determine the rate of channel widening, and identify the factors that control the widening rate. This information is necessary to understand how and at what rate alluvial channels re-attain equilibrium after a catastrophic disturbance (Graf, 1988, p. 40–42). In addition, effective management of the channel and floodplain resource requires an understanding of the present channel equilibrium (Jackson and others, 1987).

A complete understanding of the causes of entrenchment is not possible due to a lack of relevant climatic, hydrologic, and land use information. At the very least, however, entrenchment was broadly coincident with, and probably related to rapid population growth, as well as climate changes associated with the end of the Little Ice Age (Bradley, 1985). Sufficient information is available to document the magnitude and timing of channel entrenchment and widening. Results indicate that the channel and floodplain of the river developed in an entrenched, meandering-fluvial system that widened rapidly until about 1955. Channel widening since 1955 has been negligible, suggesting that channel width had adjusted to the postentrenchment conditions of discharge and sediment load. Further significant widening of the channel seems unlikely under present conditions of land use and climate. Channel stabilization evidently occurred independently of any long-term pattern of rainfall variation. Stabilization resulted primarily from development of floodplains, increased channel sinuosity, and the gradual spread of riparian vegetation into the expanding channel, a situation somewhat similar to that of the Gila River (Burkham, 1981). The decline of channel widening, however, might have been hastened by subtle climate variations and improved landuse.

Previous studies

Historic entrenchment of streams in southeast Arizona and coincident changes of range and woodland vegetation have been discussed in many research reports that include the San Pedro River. These have been summarized in several books including Hastings and Turner (1965), Cooke and Reeves (1976), Dobyns (1981), Bahre (1991), and Betancourt and Turner (1993). Interest in stream entrenchment, which is synonymous with "arroyo cutting" in the Southwest, is high among several disciplines because entrenchment is an obvious, catastrophic disruption of aquatic, floral, and surface-water resources.

Stratigraphic studies of alluvial deposits in the San Pedro and adjacent river valleys show that arroyo cutting and subsequent channel filling occurred several times during prehistoric to protohistoric time (Waters, 1985, 1988; Haynes, 1987). Although these early episodes of arroyo cutting are reasonably well known, the historic geomorphic evolution and surficial geologic history of the San Pedro River have received scant attention. Little is known, for example, about the widening process, the rate

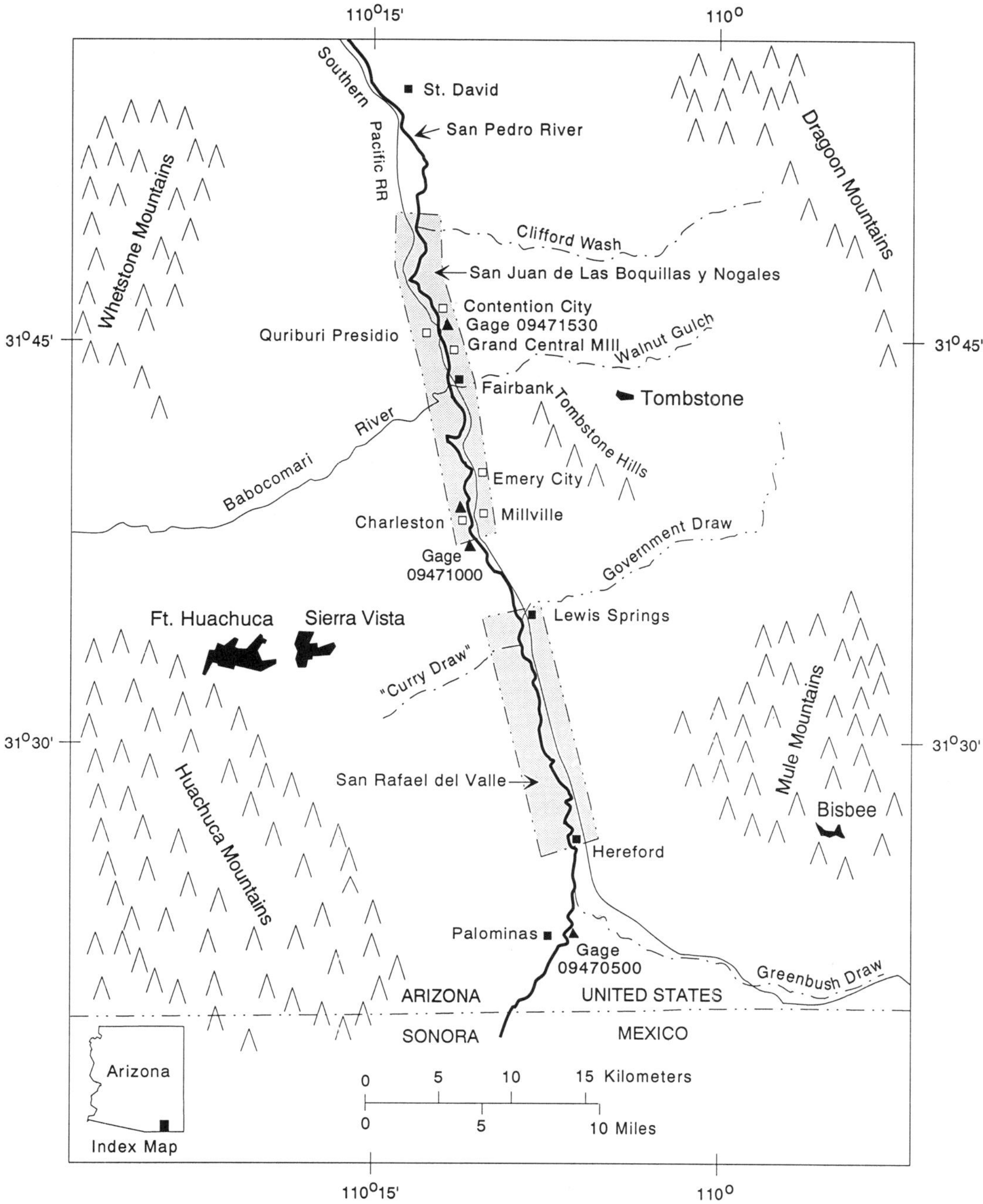

Figure 1. The study area in southeast Arizona. Field studies were undertaken in the river valley between Hereford and just north of Clifford Wash.

of channel widening, the factors that cause widening, or the time necessary to reach equilibrium.

A study by Jackson and others (1987) did address the postentrenchment channel conditions in the conservation area. Motivated by the need to understand riparian resources and identify management issues, these authors (p. 60–63) showed that the channel was evolving to a new equilibrium with present hydrologic and land-use conditions, and that understanding this equilibrium was essential to channel and floodplain management. Moreover, Jackson and others (1987) showed that the channel evolved through widening, bar development, and creation of floodplains, an evolutionary sequence similar to the descriptive entrench ment models of Elliot (1979) and Harvey and others (1985).

A study by Hendrickson and Minckley (1984) addressed the botanical history of cienegas. They showed that marshy areas were continuous along the San Pedro River before about 1890. At present, cienegas are greatly reduced in extent and modified from historic conditions. Furthermore, their results indicate that cienegas require a stable physical environment for complete development. In particular, cienegas are unstable under conditions of repeated large floods, such as occurred during the late 1800s and early 1900s. Hendrickson and Minckley concluded that concentration of cattle along water courses during the drought of 1891–1893 resulted in remarkable damage and weakening of riparian communities, eventually leading to entrenchment, although they acknowledge that a variety of causes were responsible for stream entrenchment.

Widening processes of several Southwest streams, including the San Pedro River, were studied experimentally and in the field by Meyer (1989). He determined that the rate of widening depends on the volume of bedload transported through a reach, and that the pattern (plan-form) of the entrenched channel is related to the grain size of the bedload. Meandering is enhanced through deposition of gravel-size sediment, which forces the channel to migrate around the relatively immobile gravel. This process results in development of coarse-grained point bars, a common geomorphic element of the San Pedro River channel.

Methods

This study was conducted during four seasons of field work in the winters of 1988–1991 in the Riparian National Conservation Area. Evolving river systems such as the San Pedro River leave a sedimentary and geomorphic record of their activity, even on time scales as short as a few decades (Hereford, 1986). Thus, the field studies involved mapping and interpretation of the channel and floodplain alluvium. Mapping was done on intermediate-scale (1:6,600) color-aerial photography taken in 1986 and was compiled on 1:12,000-scale topographic base maps. The age of the various channel and floodplain deposits was estimated from analysis of five sets of sequential aerial photography taken from 1937–1986. The age categories were assigned on the basis of the first appearance of a particular alluvial deposit in the aerial photographs.

In addition, the boundary of the entrenched channel was mapped using the five sets of aerial photography. The planimetrically corrected area of the entrenched channel was measured on each of five 1:24,000-scale topographic maps; this information was then used to evaluate widening in terms of channel area. Finally, historic climate data were analyzed to search for variations that might coincide with changes in the rate of channel widening.

Cultural history

The rich human history of the area, which spans nearly 11,000 yr, is relevant to understanding the relation between human activity and the evolving channel. Just outside the river valley, six Paleo-Indian sites dating from the latest Pleistocene have been extensively studied (Haynes, 1987). The sites and associated alluvial deposits are so ancient that they have little relation to the Holocene deposits of the river valley. Within historic times, the past 400 to 500 yr, the inner valley has been a locus of exploration, travel, settlement, and exploitation for three cultures—Spanish, Mexican, and Anglo (Trischka, 1971). The records of this activity, as detailed later, give an account of conditions in the valley before entrenchment.

The first Spaniard to enter the valley was a Franciscan priest, Fray Marcos de Niza, who travelled to the Zuni Pueblos in 1539, known at the time as the Seven Cities of Cibola. Only a year later, Francisco Vasquez de Coronado passed through the area during his epic journey of exploration in the Southwest. His exact route through southern Arizona is unknown, but most historians agree that the entourage traveled along the river. The group was large, consisting of about 225 horsemen, 62 soldiers, several women, and more than 1,500 Indians and camp followers, 1,000 horses, 600 pack animals, 500 cattle, and 600 sheep (Trischka, 1971; Ivey and others, 1991, p. 10–11, 41–52); these were the first domestic livestock to graze the upper San Pedro River valley. Along the river, the explorers probably encountered several villages of relatively peaceful Sobaipuri Indians, ancestors of the Pima and Tohono O'odham Indians of southern Arizona. These people were agriculturalists who grew cotton and maize using irrigated fields.

This early phase of Spanish exploration and travel was followed by the mission period of the late 17th and 18th centuries. Under the leadership of Padre Eusebio Francisco Kino, the first Christian missions and visitas, a type of mission outpost, were established in Arizona (Mattison, 1946). Two visitas, Santa Ana del Quiburi and Santa Cruz de Gaybanipitea, were established in 1692 at former Indian villages along the San Pedro River in the vicinity of Fairbank (Fig. 1; Bolton, 1936). At that time 2,000 Sobaipuri lived along the river in 15 villages. Kino visited the area several times before it was abandoned in 1701. Notable among these visitations was the trip of November 1697, when he brought 100 cattle to the Santa Cruz visita.

Kino is credited with establishing cattle ranching in the San Pedro River valley. Although he did not gain personal profit from his ranching activity, his efforts as a rancher in southern Arizona and northern Sonora show that Kino was an unusually talented businessman, which alone makes him worthy of remembrance (Bolton, 1936, p. 589). Historians regard the missionary era as the beginning of the Arizona cattle industry (Haskett, 1935). This is significant because it shows that cattle have grazed the upper San Pedro River valley in economically significant numbers for at least 300 yr.

The upper San Pedro River valley was the boundary separating usually peaceful Pima and Papago Indians or their ancestors from the fierce, war-like Apaches. Because of Apache depredations, attempts to settle the area were unsuccessful until the late 19th century. Problems with Apache Indians ended the unsuccessful 20-yr attempt to settle the valley by Kino and his associates. The final attempt at settlement by Spaniards was initiated after a treaty was concluded with the Apaches in 1768. The Quiburi area (Fig. 1) was reoccupied in 1776 as a presido garrisoned by troops transferred from northern Sonora (Kessell, 1966; Trischka, 1971). This resettlement was short-lived, however, and the area was abandoned in 1780. After 4 yr of almost continuous struggle, the Chiricahua Apaches had killed most of the troops (Kessell, 1966), forcing a retreat from the area.

After Mexican independence from Spain in 1821, settlement of the San Pedro River valley was attempted again. Basically, this was a reoccupation of the lands and sites settled by the missionaries more than a century before (Mattison, 1946). Petitions were filed with the Mexican government for land grants in the valley between 1820 and 1831. Several of these claims were eventually granted, with two of them having boundaries as indicated in Figure 1. These were cattle ranches run by a large and prosperous ranching family with numerous holdings in Arizona and Mexico. The history of these operations is poorly known; however, the Babocomari Ranch (on the Babocomari River west of the study area) had 40,000 head of cattle and large numbers of horses and mules in 1831 (Christiansen, 1988). The ranchos were abandoned by at least 1851 when the ruins of one was visited by members of the U.S. Boundary Commission (Munson, 1976).

The war with Mexico ushered in the era of Anglo exploration, settlement, and exploitation. Beginning in 1846, the records of these early explorations provide the first descriptions of the San Pedro River valley before entrenchment; they are discussed in following sections. The reports are significant because they clearly describe evidence of extensive grazing and the presence of large herds of feral horses, donkeys, and cattle. This extensive grazing, more than 40 yr before entrenchment of the San Pedro River, casts doubt on grazing as the single cause of entrenchment.

The next stage of development was Anglo settlement of the upper San Pedro River valley after the Gadsen Purchase of 1853. This development was dominated by the cattle and mining industries. Because of trouble with Apache Indians and general confusion brought about by the Civil War, significant attempts at settlement were delayed until the late 1870s. Discovery of valuable silver deposits in the vicinity of Tombstone (Fig. 1) in the late 1870s brought about rapid settlement and development of the upper San Pedro River valley (Rodgers, 1965). Tombstone flourished, and its population soared from a few prospectors in the late 1870s to 6,000 in 1881. The silver ore was processed at stamp mills built along the San Pedro River (Graham, 1976; Fulton, 1966). Five small towns sprang up around the mills; their names (shown as open squares in Fig. 1) are retained on modern maps, although they have been abandoned since the late 1800s.

The influx of miners and supporting population created a ready market for meat and vegetables. This market, completion of the transcontinental railroad through southern Arizona, and reduced threat of Indian attack heralded the era of farming and large-scale cattle ranching. Rapid settlement, depletion of woodland resources for fuel and mining timber, suppression of range fires, and reintroduction of large numbers of cattle are coincident with entrenchment of the San Pedro River (Bahre, 1991). Large-scale mining activity at Tombstone was essentially over by 1889. Since then cattle ranching and farming have been the dominant resource-based economic activity in the upper San Pedro River valley.

SURFICIAL GEOLOGY AND GEOMORPHOLOGY OF THE INNER SAN PEDRO RIVER VALLEY

The inner valley of the San Pedro River consists primarily of upper Holocene alluvial deposits, as illustrated in Figure 2. These deposits are topographically lower and have an inset relationship with the St. David Formation of Pliocene to middle Pleistocene age (Gray, 1965). The formation is lacustrine silt and marl with alluvial deposits of silt and fine sand. Pleistocene pebble- to cobble-size gravel deposits disconformably overlie the formation, forming extensive pediment surfaces in most of the area. A sequence of late Pleistocene to early Holocene deposits is exposed in several tributary streams. These deposits are notable because they contain evidence of Paleo-Indian occupation of the area 11,000 yr ago (Haynes, 1987). The late Pleistocene deposits are not present at the surface in the inner valley, except locally in the Charleston–Emery City area (Fig. 1). Evidently these deposits either were removed from the inner valley during regional early to middle Holocene erosion (Haynes, 1987; Waters, 1985, 1988), or they were buried by the upper Holocene alluvium of the inner valley. Bedrock of Cretaceous age forms the margin of the inner valley at the "narrows" downstream of Lewis Springs, between Charleston and Emery City, and from downstream of Fairbank to near the Quriburi site.

Pre-entrenchment alluvium

The inner valley consists of upper Holocene deposits that are subdivided into pre- and postentrenchment alluvium (Fig. 2). The pre-entrenchment alluvium forms a terrace that occupies most of the inner valley. Near Hereford (Fig. 1), the terrace can be subdivided into two levels separated by 0.5 to 1 m of topographic relief. The contact between the upper and lower levels cannot be traced more than 1 to 2 km, and only the lower level is present consistently in the area.

The pre-entrenchment alluvium consists of fine-grained, poorly sorted deposits of clay, silt, and fine sand with interbedded coarse sand and pebble to cobble gravel. Dark, fine-grained carbonaceous material forms one to several prominent beds in the pre-entrenchment alluvium, as shown in Figure 3. The dark beds are present throughout the area and are well exposed in the entrenched walls of tributary streams. The relatively high carbon content suggests deposition in a reducing, subaqueous environment. Thus, the carbonaceous material is the surface expression of the pre-entrenchment water table. The modern counterpart of this environment is a *cienega,* a term applied to an aquatic habitat dominated by marshy conditions, permanently saturated soils, and vegetation consisting of sedges, rushes, and grasses.

Interpretation of historic accounts suggests that cienegas were continuous in the inner San Pedro River valley before entrenchment (Hendrickson and Minckley, 1984, p. 133, 147). Stream entrenchment and lowering of the water table around the turn of the century destroyed the cienegas. Marshy, cienega-like reaches, however, are present now in the entrenched channel.

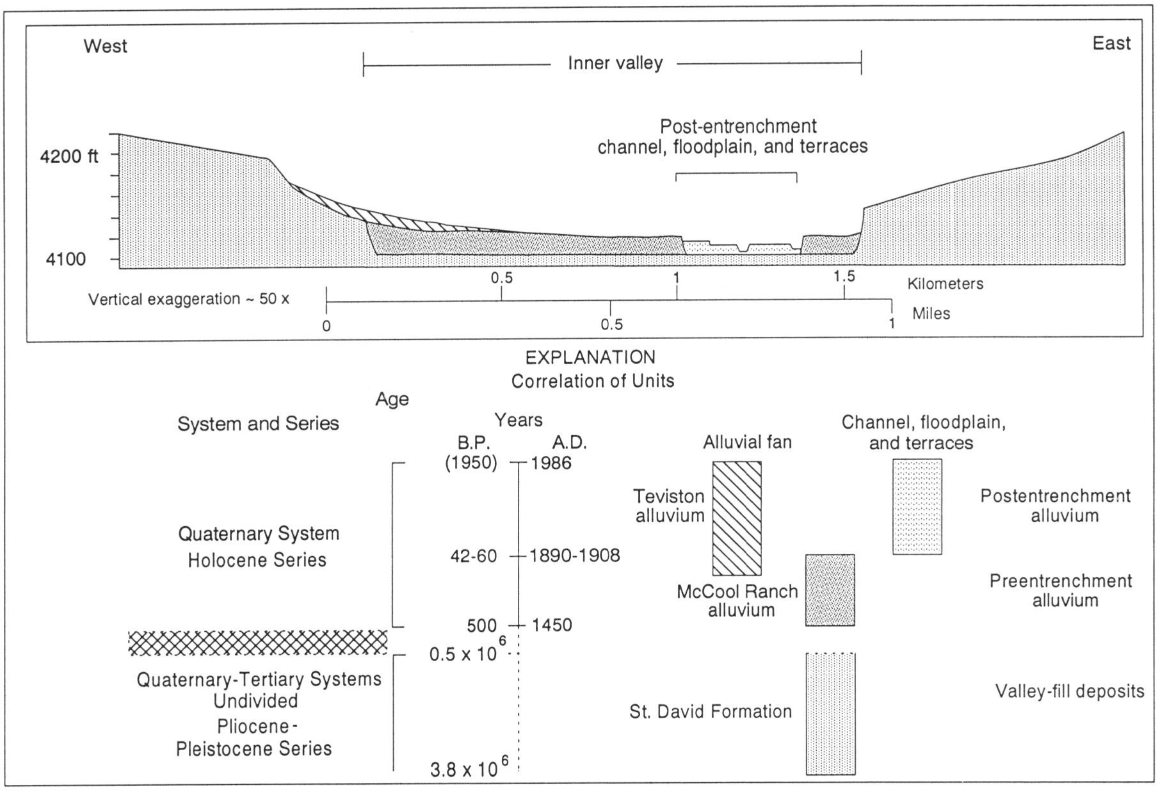

Figure 2. Geologic cross section showing correlation of surficial deposits and geomorphology of the inner valley of the San Pedro River in the vicinity of Lewis Springs. The geologic relations, geomorphology, and deposits are typical of the study area.

These reaches typically occur upstream of the junction with tributary streams. The ponding is caused by a cone of coarse sediment deposited in the channel by the tributary stream, which reduces the river gradient.

The pre-entrenchment alluvium probably correlates with the Escapule Ranch Formation of Haynes (1987). The formation is present in Curry Draw, and it can be traced into the inner valley in the vicinity of Lewis Springs (Fig. 1). As defined by Haynes (1987), the formation consists of three informally named units, which from older to younger are the "Weik Ranch, Hargis Ranch, and McCool Ranch members. The latter two probably correspond to the upper and lower terrace levels present in the inner valley near Hereford. The alluvium at the surface of the inner valley probably correlates with the McCool Ranch member. This unit records three depositional episodes, the youngest of which began about A.D. 1450 (Haynes, 1987). Deposition of the pre-entrenchment alluvium probably began by at least A.D. 1450 and lasted until entrenchment in the late 1800s to early 1900s.

In the Hereford bridge vicinity, a narrow, shallow, and sinuous abandoned channel is present on the pre-entrenchment terrace. Although poorly preserved and discontinuous, the channel is typically less than 1 m deep and only 10 to 20 m wide. Land surveys in 1901 indicate that the channel in the vicinity of Hereford varied in width from 20 to 40 m (Cooke and Reeves, 1976, p. 43). The present channel cross-cuts the abandoned channel, is substantially wider, and is entrenched several meters below the earlier channel; this was probably the channel of the San Pedro River before entrenchment.

The pre-entrenchment terrace shows evidence of local, infrequent flooding since entrenchment. This occurs between Hereford and Lewis Springs where the entrenched channel is less than 3 to 5 m deep. Near the concave bank of meanders, a recent flood overtopped the terrace, as indicated by debris containing plastic artifacts. This recent flooding did not extend more than a few tens of meters beyond the entrenched channel. An older flood, however, with well-weathered debris without plastic, extensively overtopped the terrace, producing a subdued ridge-and-swale topography of sand localized downstream of vegetation. The wide extent of the debris on the terrace suggests that it was carried by a large flood; this was probably the flood of September 28, 1926, which is the flood of record with an estimated peak discharge at the Charleston gage (Fig. 1) of 2,800 $m^3 s^{-1}$ (98,000 $ft^3 s^{-1}$).

The geomorphic effects of this flood, which is more than three times greater than the next largest flood, are not well known. The principal effect was probably deposition of gravel-size sediment. According to Meyer (1989), meandering is related to deposition of gravel-size material. Widespread deposition of

Figure 3. Photograph of a typical exposure of the pre-entrenchment alluvium 4.9 km north of Hereford. Upper Holocene Teviston alluvium overlying the McCool Ranch alluvium of Haynes (1987). Note the dark bed of cienega-type deposits near top of scale. Subhorizontal line shows contact between the alluviums. Scale divisions = 20 cm.

this coarse sediment, therefore, might have initiated meandering, as subsequent relatively low flows were forced around the coarse deposits.

Postentrenchment alluvium

From oldest to youngest, the postentrenchment alluvium consists of the terraces, floodplain, and channel of the San Pedro River, although alluvial fans have formed contemporaneously with the deposits of the entrenched channel (Fig. 2). The alluvium is inset from 1 to 10 m below the pre-entrenchment terrace. The contact between pre- and postentrenchment deposits is a distinctive topographic feature that is readily recognizable in the field (Fig. 4) and in stereoscopic small-scale aerial photographs. In transverse channel sections, the depth of entrenchment is greatest where the river is near the edge of the inner valley, because the pre-entrenchment terrace slopes up gradually toward the margin of the inner valley. In the downstream direction, the depth of entrenchment is greatest below Lewis Springs (Fig. 1), where it ranges from 5 to 10 m. Upstream of Lewis Springs, the river is entrenched only 1 to 5 m below the pre-entrenchment terrace.

Four alluvial deposits forming two terraces and two floodplains are present above the channel. In places, however, the four levels are not clearly distinguished, and only a terrace and floodplain are present. The geologic relations of the four levels on a point-bar north of Hereford are shown in Figure 5. The levels are distinguished on the basis of several factors: (1) their height relative to each other; (2) their height above the channel; (3) the size of riparian trees growing on the surface and in the alluvium; (4) the density of cryptogamic crust, a soil-protecting, crust-forming plant community consisting of algae, fungi, lichens, and mosses (Rushforth and Brotherson, 1982); (5) the content and weathering of flood debris; and (6) the topographic roughness of the surface. Generally, the younger deposits are topographically beneath older deposits, and have smaller trees, the least developed cryptogamic crust, and the roughest surface topography.

The oldest and highest terrace (unit t_1 in Fig. 5) is inset beneath the pre-entrenchment terrace (unit tm in Figs. 2, 5). The alluvium forming the terrace is typically medium- to coarse-grained sand with pebble to cobble-size gravel. At Charleston and near Lewis Springs (Fig. 1), the alluvium consists of sandy-pebble to small-cobble gravel. This gravel was probably deposited by the flood of September 1926.

Flood debris on this older terrace is well weathered and

Figure 4. Photograph showing the geomorphic and cut-and-fill stratigraphic relation of the post- and pre-entrenchment alluvium on the east side of the river 3.2 km north of the Charleston gaging station. Postentrenchment alluvium (unit t_1) forms the lower surface on the left side of the photograph. Upper surface is the pre-entrenchment McCool Ranch alluvium (unit t_m) overlain by Teviston alluvium. Note truncation of beds in the older unit. Map units discussed in text. Scale divisions = 20 cm.

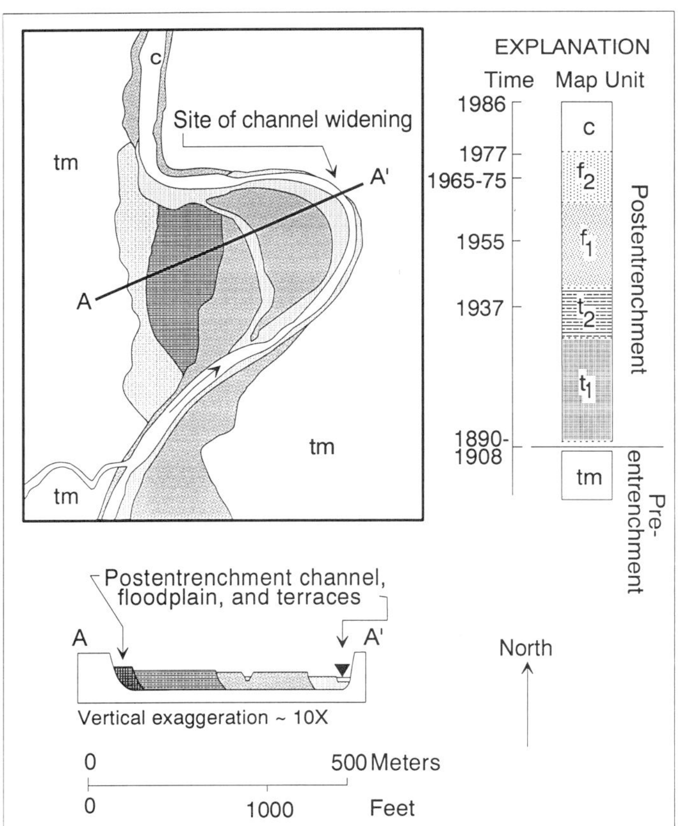

contains no plastic or aluminum, and cryptogamic crust has the densest development. Moreover, the surface typically has a relatively dense grass cover, subdued topography without ridges and swales, and large mesquite, cottonwood, or willow trees. Figure 6 illustrates the surface and vegetation of this unit near Contention (Fig. 1). The topography of the surface is flat and the mesquite have a tree-like habit. Unit t_2 of Figure 5 is broadly similar to t_1, except that it is inset as much as 0.5 m below the older unit.

The two postentrenchment terraces are present in aerial photographs of the study area taken in April 1937. Unit t_1 was vegetated and appears stabilized in the 1937 photographs, whereas unit t_2 was only slightly elevated above the channel, lacked vegetation, and appeared unstable in 1937. The lack of fresh flood debris suggests that the surfaces are not overtopped or flooded in the present channel configuration and flood regimen, and are considered terraces.

Two floodplain levels, f_1 and f_2 of Figure 5, are present in many places above the active channel. The two levels are closely related topographically, and only one floodplain is present locally. Whether f_1 should be considered a floodplain or a terrace is

Figure 5. Geologic map and cross section of the postentrenchment alluvium present on a point bar 2.4 km north of Hereford. Map units discussed in text.

Figure 6. Photograph showing postentrenchment terrace unit t_l 0.5 km southwest of Contention. The trees are mesquite. Scale divisions = 20 cm.

uncertain. In the present flood regimen, this upper floodplain is overtopped near its contact with the lower floodplain. A particularly large flood in the present channel configuration would probably overtop the upper floodplain entirely. Moreover, the youngest flood debris on f_1, as discussed below, was deposited in 1977, suggesting that the surface is within the present flood regimen. Thus, compared with units t_l and t_2, this older floodplain is not strictly a terrace.

The floodplains are inset beneath unit $t_{1,2}$ (units t_l and t_2, undivided) less than 0.5 m. Figure 7 shows the inset relation of t_l with f_1. The inset at this locality is about 50 cm and occurs along a subdued cutbank or terrace rise. Generally, vegetation and topography of this older floodplain surface differ from unit $t_{1,2}$, as shown in Figure 8. Where mesquite is the dominant tree on the younger floodplain, it has a bush-like habit rather than the tree-like habit of mesquite on $t_{1,2}$ (compare this figure with Fig. 6). The surface of the floodplain has moderately developed ridge-and-swale topography, consisting of sand deposited downstream of vegetation as a plume or ridge (Fig. 8). In addition, a moderately dense cryptogamic crust is present, as illustrated in Figure 9. This degree of development, in which about 50 to 90% of the surface area is cryptogamic crust, is typical of f_1, although development of the crust varies greatly depending on the grain size of the substrate and intensity of grazing and flooding. The effect of flooding is to scour and remove the crust, or cover it with sediment. Grazing tramples the crust retarding or precluding crustal development (Rushforth and Brotherson, 1982).

Flood debris on the two levels is similar except for degree of weathering and plastic and aluminum content. The older floodplain has less plastic and aluminum, and the material appears more weathered than debris on the younger floodplain. These characteristics vary considerably because larger floods overtop the near-channel portion of the older floodplain. A typical occurrence of flood debris on f_1 is shown in Figure 10. The debris contains very little plastic and the wood is moderately weathered compared with younger debris on f_2. This debris was probably deposited by the large flood of October 9, 1977, which had a peak discharge of 660 $m^3 s^{-1}$ (23,700 $ft^3 s^{-1}$).

The two floodplain levels have similar composition of medium- to coarse-grained sand with lenses of granule- to medium-pebble gravel. A cutbank exposure of f_1 alluvium is shown in Figure 11. The alluvium at this locality is about 2 m thick and consists of a basal gravel overlain by five beds. Each bed fines upward from a basal gravel to medium- to coarse-grained sand. Three of the sand beds are overlain by a thin, relatively dark layer of silty clay. A single overbank flood probably formed each of the three beds of sand and silty clay and each of the two sand beds. Thus, unit f_1 at this locality (Fig. 11) was deposited by at least five overbank floods. The older floodplain is not present in aerial photographs of the area taken in 1937, but it

Figure 7. Photograph showing inset relation between older postentrenchment terrace (t_1) and older floodplain (f_1) on the west side of the river 1 km south of Lewis Springs. The scale is on the lower surface; note flood debris on upstream side of cottonwood tree to right of scale. Scale divisions = 20 cm.

Figure 8. Photograph showing surface of older floodplain (unit f_1) near Contention. Bush to right of scale is mesquite. Note ridge of ridge-and-swale topography on ground left of scale. Scale divisions = 20 cm.

is present in aerial photography taken in 1955. This indicates that deposition of the older floodplain began sometime between 1937 and 1955.

Unit f_2 covers substantially less area than the older floodplain, the alluvium is thinner, and it is inset 20 to 50 cm below the older floodplain. As shown in Figure 5, the younger floodplain was deposited both along the margin of the older floodplain and in overbank channels incised into the older floodplain. Figure 12 shows a cutbank exposure of the f_2 alluvium. The alluvium is about 1.2 m thick and consists of a basal gravel overlain by medium- to coarse-grained sand with lenses of granule to pebble gravel. Three distinct beds in the alluvium are capped by thin deposits of silty clay, suggesting deposition by at least three overbank floods. The automobile tire in the alluvium is a military-

Figure 9. Photograph of moderately developed cryptogamic crust on older floodplain (unit f_1) surface 0.5 km southwest of Contention. Pebbles and cobbles to right of scale show frost-heave displacement. Scale = 10 cm.

Figure 10. Photograph showing flood debris deposited against upstream side of a willow tree on older floodplain (unit f_1). Debris is relatively weathered and contains very little plastic. Scale divisions = 20 cm.

type that was produced for several decades after World War II. The serial number indicates only that the tire was built in a year ending in seven; it could have been produced as early as 1947, but it is unknown when the tire was sold or how long it was in service before being discarded (D. W. Black, 1991, written communication). Deposition of unit f_2 was substantially later than 1947, based on the poorly constrained date of the tire. Aerial photographs suggest that the floodplain was present in 1970.

The youngest deposits are those of the channel (unit c in Figure 5). The channel contains the base flow and discharge up to bank full. A flood in March 1991 with a peak flow 60 $m^3 s^{-1}$ (2,000 $ft^3 s^{-1}$) at the Charleston gage (Fig. 1) was close to over-

Figure 11. Photograph of cutbank exposure of older floodplain alluvium (unit f_1) on west side of river 1 km south of Lewis Springs. Downstream to right. Thin dark layers are silty clay. Scale divisions = 10 cm.

Figure 12. Photograph showing cutbank exposure of younger floodplain (unit f_2) alluvium with truck tire 0.6 km south-southwest of Contention. Scale divisions = 20 cm.

topping the younger floodplain, although flows much larger than this are probably necessary to significantly inundate the f_1 floodplain. Sediment in the channel consists primarily of fine- to coarse-grained sand. The base of the channel is locally composed of pebble to cobble gravel downstream of tributary streams that carry material of this size. Bedforms in the channel consist of side bars and channel bars composed of sand that is lightly vegetated with grasses, cottonwood, willow, and saltcedar. A typical reach of the channel is shown in Figure 13. Cottonwood trees lining the channel are typically less than 5 to 10 yr old, and are partially buried by sediment.

Widening of the postentrenchment channel occurs where the channel is in contact with the pre-entrenchment alluvium, which typically occurs on the concave side of point bars (Fig. 5). At such places, little or no vegetation is present to stabilize the channel, and the steep concave bank is undercut during high runoff, particularly large floods. Reworking and undercutting of the floodplains and terraces occurs locally, although it is not widespread because the postentrenchment terrace and floodplain margins are stabilized by vegetation.

Near the margins of the inner valley, the pre-entrenchment alluvium is overlain by alluvial fans and sheetwash deposits derived from tributary streams and adjacent hillslopes (Fig. 2). The deposits consist of light colored, friable sand and gravelly sand that contrasts with the dark, cienega-type deposits of the underlying pre-entrenchment alluvium (Fig. 3). These deposits are relatively thin or not present in the axis of the inner valley. At the valley margins, however, the deposits are to 3 m thick. Typically, they have a sharp, erosional contact with the underlying McCool Ranch alluvium. Mesquite bosques, which were not present in photographs taken before entrenchment (Hastings and Turner, 1965), have subsequently developed on the surface of these sandy deposits.

Deposition of the alluvial fans and sheetwash deposits began slightly before entrenchment of the San Pedro River. The deposits are cut by the entrenched channel of the San Pedro River, suggesting that deposition began before channel entrenchment. Near the mouth of Walnut Gulch, historic artifacts dating from the turn of the century occur at the basal contact, and artifacts are present locally within the alluvium. Deposits of similar age are also present in Curry Draw (Fig. 1), where they are informally referred to as the Teviston Formation by Haynes (1987). In short, the Teviston alluvium and its correlatives in the inner valley resulted from tributary stream entrenchment and increased hillslope erosion that began entrenchment of the main channel. These entrenched tributary streams provided more efficient delivery of flood waters to the main channel, which might initiate or contribute to mainstem entrenchment.

Date of entrenchment

The postentrenchment deposits of the San Pedro River are contained entirely within the entrenched channel; most were formed after 1937, as discussed in the preceding section. Entrenchment of the channel, however, occurred at least three decades earlier. This entrenchment ended a period of alluviation that probably began about A.D. 1450 with deposition of the McCool Ranch alluvium.

The geomorphology of the pre-entrenchment channel was inferred from historic accounts and photographs, although little is recorded before 1846. Generally, these accounts suggest that the channels of the San Pedro and Babocomari Rivers were unen-

Figure 13. Photograph showing typical appearance of the channel at base flow north of Hereford. Trees are cottonwood growing on side bars. Bar surface is overtopped at discharge of 60 $m^3 s^{-1}$ (2,000 $ft^3 s^{-1}$).

trenched from about 1700 to at least 1878 (Rodgers, 1965, p. 17). According to the accounts of Kino, Manje, and Bernal in the 1690s and Velvarde in 1716, Sobaipuri Indians living on the banks of the Babocomari River and San Pedro River near Fairbank employed irrigation for agriculture, and marshy, cienega conditions existed at this time on both rivers (J. L. Betancourt, 1986, written communication).

In 1846, a battalion of about 100 military personnel traversed the inner valley along the river from near Palominas to north of St. David (Cooke, 1938). The battalion traveled with supply wagons, yet they reported no difficulty in crossing the river, a feat that would be virtually impossible with wagons if the channel was entrenched to its present depth (Rodgers, 1965, p. 16). In 1851, the banks of the Babocomari River were only 2 ft high, according to Bartlett (1854), and in 1859 the San Pedro River had a shallow bed that was almost level with the surrounding terrace (Conkling and Margaret, 1947, p. 383). A manuscript cited in Rodgers (1965, p. 17) stated that in 1875 the river was unentrenched and that a person could stoop and drink water from it at any point. Hinton (1878, p. 285) reported that the Babocomari River was a clear stream about 20 ft wide and 2 ft deep. Settlers and the domestic livestock at St. David suffered heavily from malaria in 1878, indicating marshy conditions nearby (McClintock, 1921, p. 223).

Fish were present in sufficient numbers before entrenchment to be caught by early travelers and sold commercially in Tombstone, suggesting that channel and flow conditions were much different than at present. In 1846, Cooke and his men found fish in the river up to 18 in. long, which he called salmon trout (Cooke, 1938). Bartlett (1854) supplemented scanty rations with trout caught in the Babocomari River. These fish were actually Colorado squawfish (*Ptychocheilus lucius*), formerly abundant in the Colorado River basin, but now almost extinct (Hendrickson and Minckley, 1984, p. 145). These conditions evidently persisted into the early to mid-1880s when squawfish were sold in Tombstone as buffalo fish (Gehlback, 1981), a reference to the distinct hump of the squawfish.

Written documents suggest that the channel was not entrenched until at least the early 1880s. Likewise, six photographs taken between 1882 and 1890 show that the channel was not entrenched at Contention City, Walnut Gulch, Fairbank, south of Fairbank, and Charleston (Hastings and Turner, 1965, Plates 48a, 49a, 51a, 57a; Bahre and Hutchinson, 1985, p. 183). Entrenchment probably occurred after about 1882 in the Contention City area and after about 1890 farther upstream.

The first appearance of the entrenched channel is documented by early settlers and a photograph. Cattlemen interviewed by Rodgers (1965, p. 105–107) recalled that channel cutting was completed between 1915 and 1920. However, a 1908 photograph of the flood-damaged Hereford bridge (Fig. 14) shows a channel with steep, fresh appearing cutbanks, suggesting that the channel was recently incised at this upstream location. The foregoing evidence suggests that entrenchment occurred after the period 1882–1890 in the Contention-Fairbank area and that entrenchment was completed upstream as far as Hereford bridge by 1908. Thus, entrenchment is bracketed between about 1890 and 1908, and the channel from Fairbank to Hereford, more than 32 km long, was probably entrenched in less than 18 yr.

The question of when entrenchment began is clouded by early accounts of river crossings, which indicate that the channel was incised locally long before 1890. Near St. David (Fig. 1), Bartlett (1854) reported that the banks were steep and had to be leveled before wagons could cross the river. Based on this evidence, Hendrickson and Minckley (1984, p. 147) concluded that entrenchment was local and discontinuous as early as 1850. Other evidence seems to support this interpretation: a photograph of Charleston (Fig. 1) in the early 1880s (see Hastings and Turner, 1965, Plate 51a) show a steep east-facing terrace rise, suggesting that the historic entrenchment occurred in the Charleston area by the early 1880s. Charleston, however, was built on a surface elevated several meters above the channel. Moreover, the Charleston townsite map drawn in 1879 shows a "high bank" on the east side of the site, and two streets are named West and North Terrace, respectively (Tiller, 1982).

The accounts of steep banks near St. David in the 1850s do not necessarily indicate that the channel was discontinuously entrenched at this early date. Instead, the accounts indicate the presence locally of a steep terrace rise near the river; the result of an earlier, unrelated entrenchment having an inset stratigraphic relation. Figure 15 illustrates the difference between inset and superimposed stratigraphic relations. These relations result from two or more cut-and-fill cycles in which the younger longitudinal gradient is steeper. A superimposed relation is typical of the area upstream of Lewis Springs; an inset relation occurs locally from Charleston downstream to the northern end of study area. Here, early travelers had difficulty crossing the river and had to level steep banks, which were actually a terrace rise. For these people, concerned with crossing a hostile and dangerous wilderness, the distinction between inset and superimposed relations was irrelevant. What they described as a river bank was probably a terrace rise.

Cause of entrenchment

A series of large floods in the late 1800s to early 1900s was the immediate cause of entrenchment. These floods were the agents of erosion that deepened and widened the channel of the San Pedro River. Other studies of Southwestern streams have found that historic entrenchment and subsequent widening were associated with a high frequency of large floods (Burkham, 1981; Hereford, 1984, 1986; Webb, 1985; Betancourt, 1990). The difficult question about which much controversy swirls is the cause of the floods. A vast literature has developed regarding the causes of historic stream entrenchment in the southwest United States (see reviews in Cooke and Reeves, 1976; Graf, 1983; Bahre, 1991; Betancourt and Turner, 1993). For the most part, the proposed explanations revolve around climate change and land use, or some combination of the two.

Within the study area, diastrophism was probably an addi-

Figure 14. Photograph of the Hereford bridge in 1908 showing entrenched channel. View is upstream. Photograph from Bisbee Mining and Historical Museum.

tional factor in hydrologic change. On May 3, 1887, the San Pedro River valley was rocked by a major earthquake, which, excluding California, is the largest seismic event recorded in the western United States (DuBois and Smith, 1980). The epicenter was near Batepito in far northeastern Sonora, about 50 to 75 km southeast of the upper San Pedro River valley, with an estimated magnitude of 7.2. Seismic intensity in the study area was VIII-XI on the Modified Mercalli Scale of Intensity, based on accounts of cracks, gaping fissures, subsidence, and damage to structures (DuBois and Smith, 1980, p. 58). Documented hydrologic effects of the earthquake include development of a fissured zone the length of the valley and changes in the water table and streamflow. These events preceded stream entrenchment by at least 3 yr; nonetheless, disruption and lowering of the water table might have preconditioned the channel system for the rapid flood-induced entrenchment of the period 1890–1908.

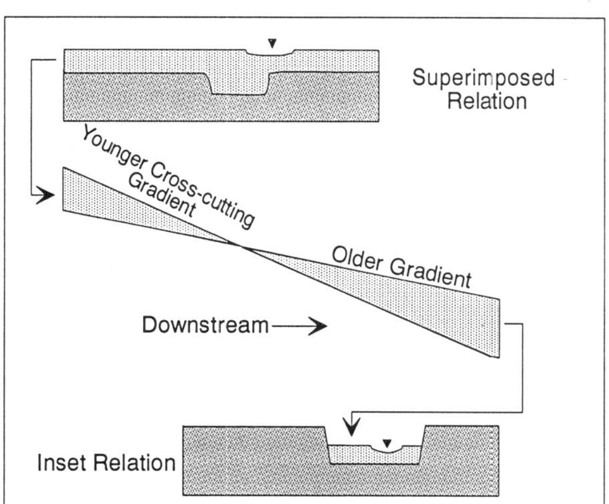

Figure 15. Superimposed and inset stratigraphic relations and their geomorphic expression.

Using newspaper accounts and other documents, J. L. Betancourt (written communication, 1986) developed a flood history for the late 1800s to early 1900s. Floods began to draw attention in 1881 when a dam upstream of Charleston (Hastings and Turner, 1965, Plate 49a) washed out, and the channel at Charleston was widened and deepened. According to Hastings and Turner (1965, p. 158), this dam was ultimately destroyed by floods in 1887. Damaging floods were reported in local newspapers in July, August, and September of 1887. The largest floods in several years again caused damage in the upper San Pedro valley in August 1890. In August 1891, floods caused extensive damage to farms and the railroad through the upper valley. These floods were preceded by high runoff and some flood damage in March of that year. A large flood in August 1893 threatened Fairbank and stalled railroad traffic south from Benson. The following August again produced large floods that washed out a dam at St. David and damaged ranches along the river. Newsworthy floods evidently did not occur in 1895, but in 1896 extensive flood-related damage was reported in July, August, September, and October. A 4-yr hiatus occurred until September 1900, when flood-weakened bridges delayed trains. August of 1901 brought troublesome floods to the lower San Pedro that presumably affected the study area. Finally, floods in

February and August 1904 and in January and March 1905 damaged structures and shifted the channel locally.

These floods were clearly associated in time with channel entrenchment, although it is unknown if floods of this frequency and size were typical before entrenchment. This seems unlikely, however, given the unincised morphology of the pre-entrenchment channel; these floods were probably unusual. Moreover, the pattern probably persisted until the 1940s to early 1950s and was associated with the postentrenchment widening discussed in a following section. Betancourt (1986, written communication) does not mention floods between 1905 and 1915, but the published records from the stream gage near Charleston (Fig. 1), which began operation in October 1915, indicate that large floods were typical of the period 1915 to the 1940–1950s.

The cause of the presumed increase in the frequency of large floods that led to entrenchment is not well understood. Land use is cited by ecologists and social scientists as the principal cause of most historic changes of vegetation and fluvial systems in southeast Arizona, whereas physical scientists cite climate change independently or in association with human activity as the principal causative factor (Bahre, 1991, p. 41–58). Evidence in support of one interpretation over another is conflicting or lacking.

Overgrazing and related activity associated with Anglo settlement of the region are appealing causes, but large numbers of cattle have been present in the upper San Pedro River valley since at least 1820. Cattle were introduced into the valley in 1697, if not a decade earlier (Bolton, 1936; Trischka, 1971), during the Spanish-Mexican phase of the Arizona cattle industry (Haskett, 1935). Following this early activity, the valley was settled and cattle ranching was undertaken during the period 1820–1831 when petitions were filed by Mexican Nationals for the land grants (Mattison, 1946) forming the present Riparian National Conservation Area (Fig. 1). These operations were unsuccessful because of Apache depredations, and the ranches were soon abandoned. Although ranching was unsuccessful at this time, the abandoned livestock evidently multiplied successfully without human intervention.

Cook (1938) found numerous feral cattle, remnants of the domestic herds of the early settlers, during a traverse of the inner valley in 1846. He reported that traces of cattle were as abundant as buffalo signs on the Great Plains. These included cattle and horse trails as numerous as those in Missouri, which at that time had been settled for 20 to 40 yr. In the vicinity of the Babocomari River, Bartlett (1854) visited an abandoned ranch and learned that 40,000 head of cattle plus large numbers of horses and mules had grazed in the area. The largest recorded cattle population of the upper San Pedro River valley was 36,000 in 1890, according to the Cochise County tax rolls, which reported the assessment plus 50% (Rodgers, 1965, p. 68). Thus, the cattle population of the early 1800s to mid-1800s was possibly as large as the highest levels of the late 1800s.

Historic references to large herds of cattle during the early to mid-1800s are quite specific, particularly within the upper San Pedro River valley. Christiansen (1988) estimated that as many as 100,000 animals were grazing in the valley and adjacent areas, based on reports of early explorers and assuming that the ranchers abandoned substantial numbers of cattle that reproduced successfully. Similar estimates were made by Hastings and Turner (1965, p. 34), who reported the number of wild cattle at 50,000 to 100,000. Bahre (1991, p. 114–115), however, maintained that the number of cattle reported was impossibly large because the lack of developed water sources precluded large numbers of cattle.

Developed water sources, however, were probably unnecessary in the early to mid-1800s when the domestic water requirements of the small human population were negligible. Even if it was necessary to develop water sources, evidence of this activity dating from the Mexican ranching period has not been found. By 1881, however, the population of the upper San Pedro River valley was about 6,000 (Rodgers, 1965, p. 53). At this time, most of the easily obtainable surface water was probably used for domestic and mining purposes. Additional water for livestock would require development of wells, stock ponds, and other water gathering or retention devices.

Thus, accounts of early Anglo explorers and travelers are probably correct, or at the very least cannot be refuted. Large numbers of feral livestock and evidence of grazing were probably typical of the upper San Pedro River valley in the early to mid-1800s. Therefore, the role of grazing in stream entrenchment and watershed adjustment around the turn of the century is not clear because grazing preceded entrenchment by more than 40 yr.

Likewise, the role of climate change is uncertain. The increased flood frequency during and following entrenchment has not been related to specific rainfall changes in the upper San Pedro River drainage basin. In the Santa Cruz River drainage basin, however, Betancourt and Turner (1993) show that July–August rainfall at Tucson was unusually high during entrenchment. Moreover, the frequency of high-intensity rainfall was large, while the frequency of low-intensity rains was low. These factors suggest that the large floods associated with entrenchment of the Santa Cruz River, which was roughly contemporaneous with entrenchment of the San Pedro River, were generated by unusually high rainfall—rainfall with few analogies in the 20th century. The climate data are from a single station that has been moved many times in its long history. Thus, regional application and validity of the data are questionable.

Unusually intense rainfall at Tucson during entrenchment of the Santa Cruz and San Pedro Rivers was associated with strong and frequent ENSO (El Niño Southern Oscillation) events (Betancourt and Turner, 1993), suggesting that it was a regional phenomenon. A complex system of global climate fluctuations, ENSO typically increases rainfall in the Southwest during spring and late summer to fall (Ropelewski and Halpert, 1986; Andrade and Sellers, 1988). This enhancement of July-August rainfall during the late 1800s by ENSO is unusual and has not been typical of the 20th century. Nonetheless, it seems likely that increased summer rainfall was regional and that it also influenced conditions in the San Pedro River valley. Betancourt and Turner

(1993) point out that the intensified ENSO activity preceding and during entrenchment was probably related to long-term climate change associated with the global warming at the end of the Little Ice Age.

In short, climate at the end of the 19th century might have differed from the climate of the previous 300 yr, when the pre-entrenchment alluvium was deposited. Entrenchment, therefore, was the response of an alluvial system that was not in equilibrium with the unusual rainfall and floods of the late 19th century. The response of the system, however, might have been less catastrophic in the absence of human activity. Overgrazing, suppression of range and woodland fire, and extensive wood cutting in nearby mountains (Bahre, 1991) were factors that exacerbated entrenchment, but whether these activities were the ultimate cause of entrenchment is unknown.

Overgrazing is frequently cited as one of the fundamental causes of entrenchment (Bahre, 1991), and the poor range conditions of the late 1800s are usually taken as evidence of overgrazing (Tourmey, 1891; Thornber, 1910). Yet studies of range hydrology show that estimates of range condition are generally inadequate to evaluate the hydrologic effect of grazing (Gifford and Hawkins, 1978). Moreover, the obvious detrimental effects of grazing are localized around stock tanks and fence lines (Hereford, 1984), and in relatively large basins climate effects generally overwhelm those of grazing (Graf, 1988, p. 241). Thus, the poor range conditions of the late 1800s could have resulted from adverse climate factors as well as from grazing.

Perhaps the greatest limitation to understanding the specific link between climate and entrenchment in the upper San Pedro River valley is the paucity of data. The kind of detailed weather and runoff data needed to assess subtle changes in rainfall and runoff are scarce to nonexistent. Published reliable discharge measurements have been recorded only since 1916, and the earliest available daily weather records began in 1895. Thus, pre-entrenchment climate and runoff of the area are unknown.

CHANGE IN RIPARIAN FOREST OF THE INNER VALLEY

In terms of channel stabilization and floodplain development, the riparian forest is probably the most important floral element of the inner valley. The extensive root systems resist fluvial erosion, and the trunks and branches slow the velocity of flood water, causing sediment deposition. Historic vegetation change in the San Pedro River valley uplands has been extensively studied (see the review in Bahre, 1991), but development of the inner valley postentrenchment riparian forest is not well known. Several workers have noted that native riparian trees have generally increased in southeast Arizona since the turn of the century (Hastings and Turner, 1965; Bahre and Bradbury, 1978; Gehlbach, 1981). Jackson and others (1987) specifically noted the development of a riparian forest within the study area. During this investigation, it was found that development of the present riparian forest is closely related to expansion of the entrenched channel and deposition of floodplain alluvium. Expansion of the forest is documented in sequential-aerial photography of the area from 1937 through 1986, and in ground-based photography from 1882 to the 1930s.

Type and spatial distribution of the riparian forest

The present riparian forest is composed mainly of winter deciduous broadleaf trees. The native trees are cottonwood (*Populus fremontii*), gooding willow (*Salix gooddingii*), seepwillow (*Baccharis glutinosa*), and mesquite (*Proposis juliflora*). Saltcedar (*Tamarix chinensis* Lour.) is a nonnative bush or tree that locally forms dense groves. These plants are phreatophytes with taproots connected directly to groundwater (Graf, 1988, p. 248–250). The spatial distribution of the trees varies along the axis of the valley, as well as in transverse cross sections. Downstream from Hereford to near Lewis Springs, the floodplain and terrace (units $f_{1,2}$ and $t_{1,2}$, respectively) are dominated by cottonwood and willow. In this area, mesquite and saltcedar are rare to absent in the entrenched channel. In transverse sections, mesquite forests occur mainly outside of the entrenched channel on sandy deposits of the Teviston alluvium that cover the pre-entrenchment terrace. Saltcedar increases in abundance downstream of Lewis Springs, and at the northern boundary of the area is the dominant vegetation of the entrenched channel, although cottonwood and willow are also present. Mesquite is also present in the entrenched channel downstream from Lewis Springs, and locally, such as near Contention (see Figs. 6, 8), it is the dominant vegetation.

Pre-entrenchment vegetation of the inner valley

Riparian trees in dense forests were not abundant before entrenchment, as indicated by historic accounts from the mid-1800s and from photographs of the late-1800s. The lack of a dense riparian growth might be the result of fuelwood cutting (Bahre and Hutchinson, 1985); however, this paucity evidently preceded Anglo settlement, which began about 1870 (Rodgers, 1965). Moreover, few cut stumps of cottonwood or willow were found outside the entrenched channel, suggesting that either the stumps were not preserved or that few trees were present to cut.

A rather clear reconstruction of vegetation density in the study area in 1846 can be inferred from the accounts of the "Mormon Battalion" (Cooke, 1938). Led by Lt. Colonel Philip St. George Cooke, the battalion consisted of about 100 men, horses, and a number of supply wagons. Cooke was charged with building a wagon road from Santa Fe, New Mexico, to the Pacific Coast as a supply route for troops fighting in the war with Mexico (Christiansen, 1983). The battalion traversed the inner valley of the San Pedro River downstream from Greenbush Draw to north of St. David (Fig. 1) in December 1846. The road builders traveled along the river, except for short diversions around the west side of the three narrows, near Lewis Springs, Charleston, and upstream of Contention, which Cooke (1938) referred to as canyons. Although Cooke (1938) mentioned mesquite and "ash," these evidently were not dense enough to be a problem for transportation, even for a group encumbered with wagons. Figure 16 shows the area downstream of the Charleston narrows traversed by Cooke on December 11 and 12, 1846. The dark vegetation on either side of the channel is a mesquite forest that at present is difficult to penetrate, even on foot. These dense mesquite forests probably developed since 1846.

The pre-entrenchment terrace near Contention (Fig. 1) is presently covered by a dense mesquite forest; the postentrenchment channel is covered by a dense growth of mesquite, saltcedar, and willow. A photograph of this area taken in 1882 (Bahre and Hutchinson, 1985) shows no cottonwood or large trees, and a relatively low density of small trees, presumably mesquite. Likewise, photographs of the inner valley taken in the early 1890s near Fairbank, Charleston, and Millville lacked a riparian forest of any variety (Hastings and Turner, 1965, p. 156–174). Presently, a dense mesquite forest covers the pre-entrenchment terrace in these areas, and the postentrenchment channel has a well-developed forest of cottonwood and willow.

Development of postentrenchment riparian forest

The riparian forest of the postentrenchment channel did not develop until after the late 1930s. From entrenchment until at least the late 1930s, the entrenched channel lacked a significant density of riparian trees. For example, a photograph of the entrenched channel at the Hereford bridge in 1908 (Fig. 14) shows no trees in the channel; presently, the dense cottonwood forest at this locality obscures the site of the early photograph. Figure 17 shows the changes in riparian vegetation at the Palominas bridge between 1939 (Fig. 17A) and 1991 (Fig. 17B): trees were not abundant in 1939, but cottonwood and willow now form a dense forest in the entrenched channel. Likewise, at the stream gage downstream of Charleston (Fig. 1), trees were virtually absent in the entrenched channel in 1930 (Fig. 18A), a situation that has since been completely reversed, as shown in Figure 18B.

By 1937, aerial photographs show that cottonwood and presumably willow were present locally, although the density of riparian trees in the entrenched channel in 1937 was much below levels in subsequent aerial photographs. Many of the trees present in 1937 are identifiable in the field where they occupy the older terrace of the entrenched channel, as previously discussed. Figure 19 shows a cottonwood grove south of Lewis Springs that was not present in the 1937 aerial photographs. These trees are growing in an entrenched meander on and within f_1 alluvium deposited between 1937 and 1955. They have the mature growth habit and size typical of cottonwood trees germinating during this period.

In summary, the density of riparian trees in the inner valley has increased significantly since the mid-1800s, if not entirely

Figure 16. Downstream view of the inner valley showing entrenched channel of the San Pedro River and dense mesquite forest on terrace on both sides of river, Southern Pacific Railroad in foreground. After passing around the narrows downstream of Charleston, the Mormon Battalion (see text) reentered the inner valley near here and proceeded downstream along the river.

since entrenchment around 1890–1908. The dense mesquite forest developed on the pre-entrenchment terrace was evidently not present in 1846. Moreover, it was probably not present in the late 1800s. The cottonwood, willow, and saltcedar forest of the entrenched channel became established, for the most part, after 1937. The lack of dense riparian forests before entrenchment was probably the result of the high water table and marshy conditions associated with the widespread cienegas that were typical of the pre-entrenchment era. After entrenchment, the lowered water table and expansion of the channel provided recruitment sites for subsequent forest development in the entrenched channel. Rapid development of the forest after about 1937 probably resulted from less frequent large floods and the increased width of the channel. As the channel widened, recruitment sites became larger

Figure 17. Palominas bridge; view is upstream. (A) Photograph taken in 1939 (from Special Collections, University of Arizona, Tucson); (B) 1991. Dense riparian vegetation has developed in the channel.

Figure 18. River channel at the unnumbered gaging station downstream of Charleston. View is downstream. (A) Photograph taken on April 17, 1930 (from files of the U.S. Geological Survey, Tucson, Arizona). (B) Approximately the same scene in 1991; the camera is closer and has a longer focal length.

and more abundant; a critical width was probably reached by the 1930s that provided space on relatively stable surfaces for rapid expansion of the riparian forest. On the pre-entrenchment surface, mesquite recruitment was probably enhanced by the lowered water table and deposition of the sandy Teviston alluvium.

RATE OF CHANNEL ENLARGEMENT

The spatial distribution and the age of the postentrenchment alluvium (Fig. 5) indicate clearly that the area of the channel and floodplain have enlarged since initial entrenchment around the

Figure 19. Photograph of cottonwood trees in an entrenched meander 4.1 km south of Lewis Springs. Note automobile parked on pre-entrenchment terrace. These trees are less than 54 yr old, as the meander scar and the trees were not present in 1937 aerial photographs. Scale divisions = 20 cm.

turn of the century. In an alluvial system with a strong component of lateral migration such as the San Pedro River, progressively younger floodplains form as the channel migrates. Migration of the channel simultaneously erodes the pre-entrenchment alluvium, while providing space for subsequent floodplain deposition. In this fashion, the riparian habitat increased in size. Two important questions emerge about this evolutionary process: what is the rate of channel widening and is the evolutionary process complete?

Answers to these questions are important for management of the riparian resource. For example, if the process is complete, the system might shift from lateral to vertical accretion, the area occupied by floodplains will no longer increase, and the channel could eventually fill with sediment, possibly causing a significant change of riparian habitat. Ultimately, the channel could attain pre-entrenchment conditions, which might include extensive development of cienegas. On the other hand, if the channel is still widening, then floodplain area should increase with continued expansion of the present riparian habitat. From a theoretical point of view, the time scale of widening is unknown and the factors driving widening are poorly understood. Although the channel should adjust or attain a new equilibrium with postentrenchment flow conditions, the time necessary to reach this equilibrium is unknown. Moreover, external factors such as climate and land use will hasten or delay the time necessary for adjustment.

Method

The rate of channel enlargement was estimated by determining channel area through time. Channel area is defined as the area between the walls of the entrenched channel for a specified length of channel at a point in time. The entrenched channel was mapped on sequential, stereoscopic small-scale aerial photography. As previously discussed, the channel walls are identifiable in stereoscopic aerial photographs because the walls form a nearly vertical, continuous feature that separates two broad surfaces of different elevation. The mapped channel boundary was compiled on 1:24,000-scale topographic maps, thereby rectifying the distortion of the aerial photographs. Photographs from five surveys taken between 1937 and 1986 were obtained that cover the entire area from Hereford bridge to 3.1 km downstream of Contention (Fig. 1). The date, source, and scale of the five surveys are listed in Table 1. Channel area was measured using digital methods on each of five topographic maps of the study area for each of the five surveys.

Results

Figure 20 illustrates expansion of the channel from pre-entrenchment to 1986 in a 2-km reach of the river beginning 3.2 km downstream of the Hereford bridge (Fig. 1). This area was chosen because the pre-entrenchment channel is recognizable in

TABLE 1. AERIAL PHOTOGRAPHIC SURVEYS HAVING FULL STEREOSCOPIC COVERAGE OF THE STUDY AREA* USED TO MAP CHANNEL ENLARGEMENT AND SURFICIAL GEOLOGY

Date	Source	Scale
April 1937	SCS	1:30,000
January 19, 1955	USGS	1:20,000
January 26, 1955	USGS	1:20,000
October 10, 1970	USAF	1:55,000
October 12, 1978	SCS	1:25,000
October 13, 1978	SCS	1:25,000
September 11, 1986	BLM	1:6,600

*See Figure 1.

the field, and it was measured near here by surveyors in 1901 (Cooke and Reeves, 1976, p. 43). The depth of the abandoned channel has increased in its lower half through diversion of tributary runoff into the channel by construction of the railroad, which was completed in its present alignment (Fig. 1) in December 1903 (Myrick, 1975, p. 208). This diversion, however, has not affected the width of the channel because the width is similar upstream and downstream of the input point.

The channel in 1937 at this locality has little geomorphic relation to the earlier channel; indeed, this younger channel cross-cuts the abandoned channel. Initial entrenchment probably resulted first in a major realignment of the earlier channel, which was followed by widening of the realigned channel. In contrast, the post-entrenchment channel has progressively widened within the same alignment (Fig. 20). Widening evidently occurred by lateral migration and expansion of entrenched meanders. For example, only one meander scar was present in 1937 at the southern end of the reach. In 1986, this scar had expanded substantially and the number of scars in the reach increased to six.

Channel area increased rapidly from entrenchment to 1955; since then the rate of increase of channel area has declined. The reconstructed area of the pre-entrenchment channel was about 6.9 hm^2 (Table 2). This estimate is based on the area per unit channel length (8.77 hm^2 km^{-1}) referenced to the length of the channel in 1937. Between 1900 and 1937, channel area increased by a factor of 2.4 to 16.4 hm^2. From 1937–1955, channel area almost doubled again to 32.6 hm^2; from 1955–1986, channel area increased from 32.6 to 39.2 hm^2, a factor of only 1.2. The 9.7% change in area between 1978 and 1986 (Table 2) suggests a reversal of this pattern. The reversal is a local anomaly that is not evident in longer channel segments.

The spatial variation of channel enlargement is shown in Figure 21, which illustrates the cumulative postentrenchment channel area in four reaches of unequal length. The four reaches (shown in Fig. 21) correspond to four of the five 1:24,000-scale topographic sheets of the study area. In the downstream direction, these are the Hereford, Nicksville, Lewis Springs, Fairbank, and Land Quadrangles, respectively. The Nicksville Quadrangle was excluded from the figure because channel area is too small to show. The pattern of change in each quadrangle is generally the same, except for the Land Quadrangle, in which enlargement from 1955–1970 and from 1978–1986 was negligible. In the three upstream reaches, the channel area increased substantially between 1937 and 1955. In contrast, an increase of channel area from 1955–1986 ranged from only about 40 to 60% of the earlier

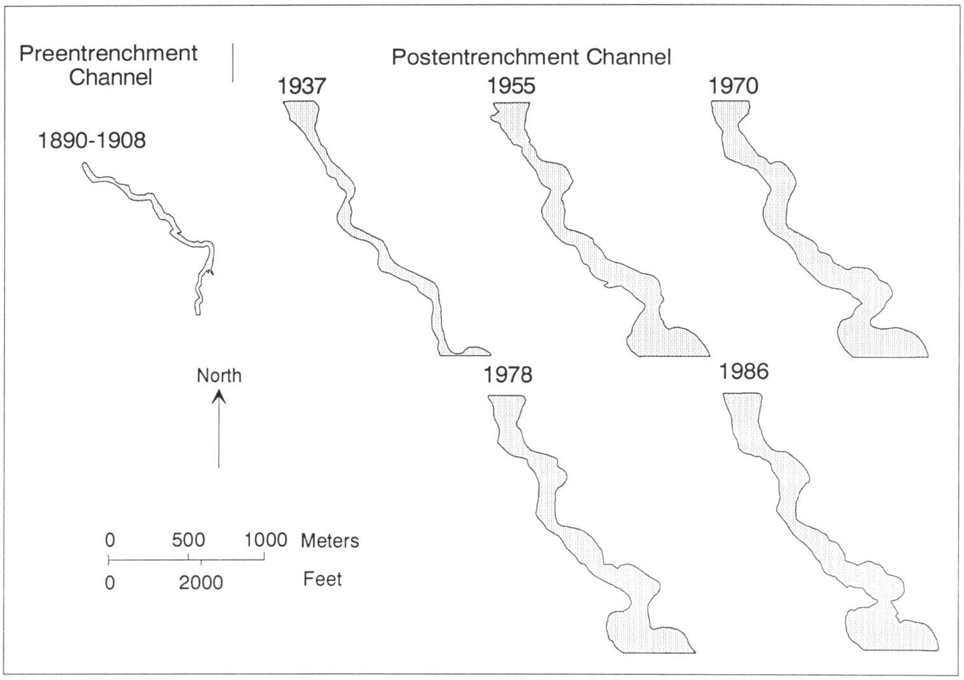

Figure 20. Maps showing the pre-entrenchment channel and expansion of the postentrenchment channel 3.2 km north of Hereford bridge as compiled from sequential-aerial photography.

TABLE 2. INCREASE OF CHANNEL AREA OF A 2-KM REACH NORTH OF HEREFORD*

Year	Acres	Area (hm^2)	Increase (%)
1890-1908	17.1	6.9	....
1937	40.6	16.4	58
1955	80.5	32.6	49.5
1970	86.3	34.9	6.7
1978	87.6	35.4	1.5
1986	97.0	39.2	9.7

*See Figure 20.

period, even though the later period is 13 yr longer than the preceding period. In short, the pattern of substantially reduced channel enlargement since about 1955 is probably typical of the entire study area, wherever the channel is free to enlarge.

The temporal variation of channel enlargement is shown in Figure 22. This figure shows the measured cumulative area of the entrenched channel deposits as a function of time, as well as the estimated area of the pre-entrenchment channel. The reconstructed area of the pre-entrenchment channel is 1.3 km^2, a figure whose accuracy is unknown. Nevertheless, it appears to be a reasonable extrapolation of the measured data. Considering the entire study area and assuming that entrenchment occurred by 1900, the estimated rate of enlargement from 1900–1955 was 0.109 $km^2 yr^{-1}$, and from 1956–1986 the rate was only 0.024 $km^2 yr^{-1}$. The rate of enlargement, therefore, has been small for at least 31 yr. The conclusion is that the rate of channel enlargement has declined in recent years. Furthermore, this probably signifies stabilization of the channel and the end of significant widening.

CLIMATE, RUNOFF, AND CHANNEL WIDENING

Entrenchment and subsequent stabilization of the San Pedro River channel are closely related to flood history. Initial entrenchment at the turn of the century resulted from the large floods of that era. Likewise, the continued expansion of the channel until at least 1955 was probably a continuation of the entrenchment process driven by large floods. Figure 23 shows the annual flood series of the San Pedro River at the Charleston gage (Fig. 1) from 1916–1987. The data reveal that large floods were

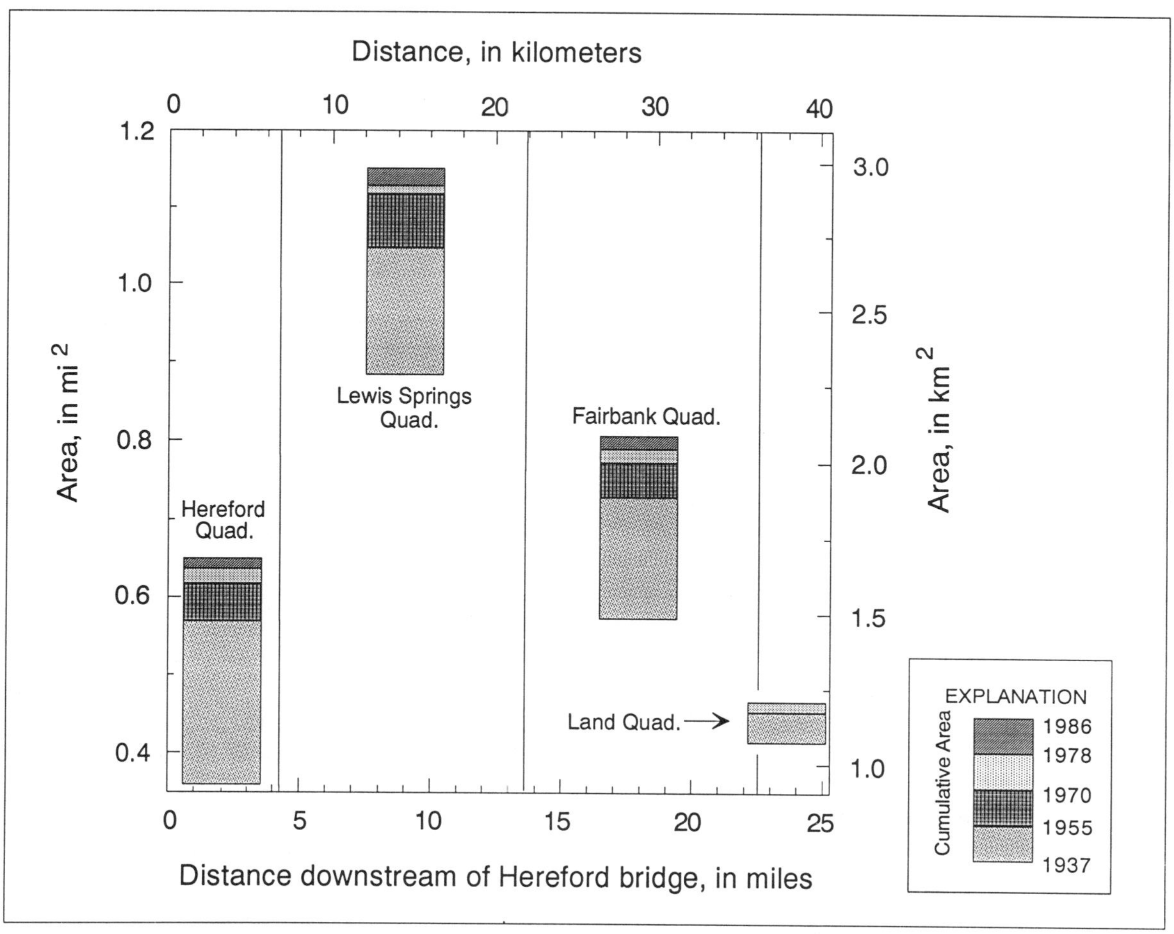

Figure 21. Spatial variation of channel enlargement downstream from the Hereford bridge. Vertical lines show length of studied reaches within the four 1:24,000-scale topographic maps.

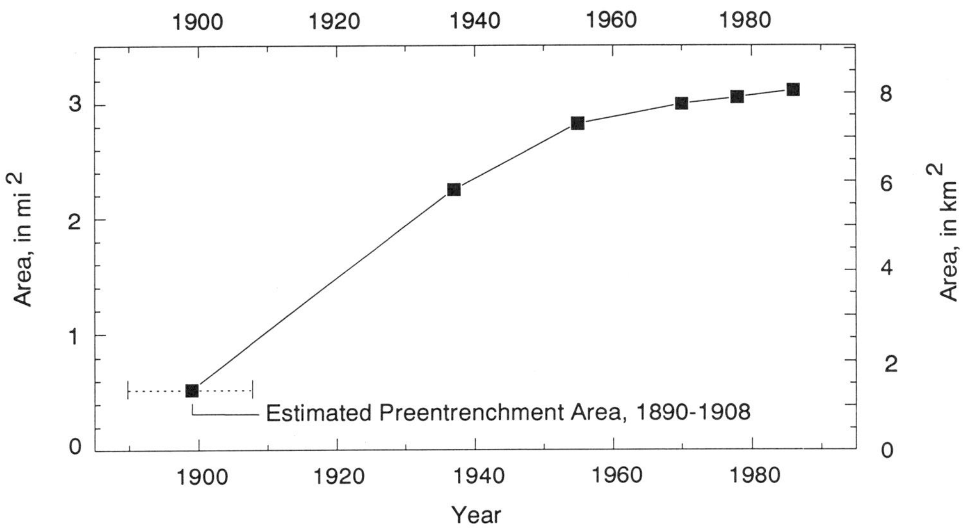

Figure 22. Time series showing cumulative area of the entrenched channel. Dotted horizontal line shows time of entrenchment from 1890–1908. Channel expansion slowed appreciably by at least 1955.

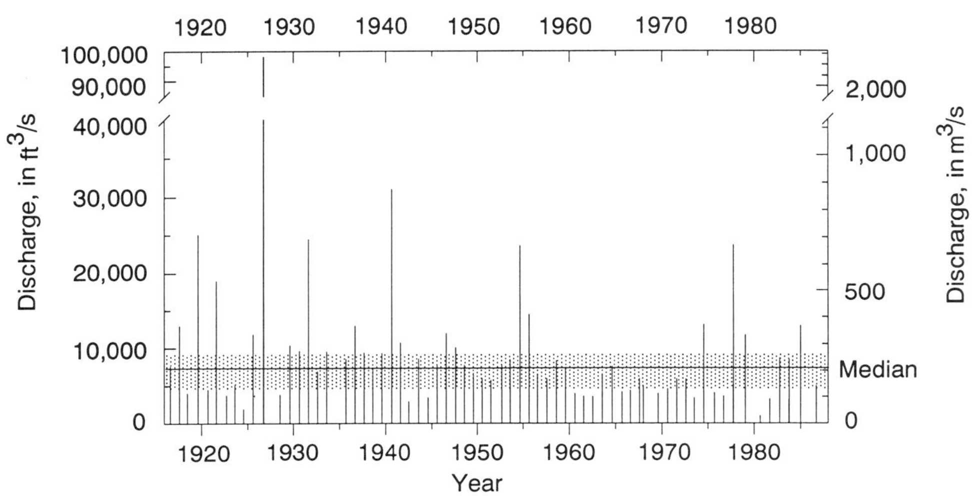

Figure 23. Time series showing annual flood of the San Pedro River at Charleston (Fig. 1), 1916–1987. Pattern shows the 25th through 75th percentiles.

relatively frequent during about the first half of the record. Seventeen floods equal to or greater than the 75th percentile of all annual floods occurred between 1916 and 1955, an average rate of about one large flood every 2.4 yr. This period includes the flood of record, which occurred on September 28, 1926 (Fig. 23). In contrast, from 1956–1987, only four floods were in the upper quartile, an average rate of one large flood about every 8 yr. Only one of these floods, the flood of October 9, 1977, was comparable in size to the largest floods of the earlier period (Fig. 23).

Peak-flood discharge was probably modified or attenuated by the morphology of the evolving channel. Increased sinuosity of the channel and development of floodplain vegetation would produce a reservoir effect, thereby reducing peak-flow rates (Burkham, 1976, 1981). In addition, infiltration of flood waters or transmission losses (Lane, 1990) would increase over time as the channel widened, which would also reduce runoff volume and peak-flow rates. Thus, the long-term pattern of reduced peak-flood discharge is related partly to channel widening, and might be independent of other factors. Nevertheless, because climate and land use affect the amount and frequency of water delivery to the channel, they probably control the time necessary for channel sinuosity and vegetation to change flow rates.

This section analyzes climate and runoff. The analysis is a search for a link between climate and entrenchment and for rainfall variations that might explain the reduced frequency of large floods after 1955. Toward the latter objective, the analysis identifies flood and rainfall seasonality, examines the relation between antecedent rainfall and floods, analyzes seasonal variability of rainfall, and develops time series of rainfall intensity.

Data and methods

Daily precipitation from eight weather stations in or near the study area were used to evaluate historic climate variations. The location, elevation, and period of record of the eight stations are listed in Table 3. The data consist of 24-hr rainfall measurements collected mainly at cooperative stations staffed by volunteers (NOAA, 1986). The data were obtained on magnetic tape from the National Oceanic and Atmospheric Administration (NOAA), Asheville, North Carolina. Monthly precipitation summaries of these stations are in Green and Sellers (1964) and Sellers and Hill (1974). The complete weather data set consists of 212,203 observation days of which 178,257 observations were actually made and recorded.

The number of missing entries varies substantially among the weather stations (Table 3). Apache Powder Co., Benson, Patagonia, and Tombstone have reasonably complete records. The missing entries typically range from several days to several months at most. The remaining stations have substantial gaps of several years' duration. The hydrologic data also have missing entries (Table 3), primarily in the early part of the record in which several years are missing.

In the following analysis, a missing value was assigned to a season if more than 10% of the daily entries of a particular station were absent. The number of stations reporting annually varies because of missing entries and an increase in the number of stations over time. Figure 24 shows the number of stations reporting annually from 1895–1987 for the wet season, June 15–October 15. The number ranges from one to three stations reporting before 1900, from four to six, between 1901–1976, and from three to four from 1977–1987.

Seasonality of rainfall and runoff

The annual precipitation cycle at Tombstone is shown in Figure 25. The cycle has a distinct wet season from mid-June through mid-October or early November, a pattern typical of all weather stations in this region (Table 3). Operationally, the wet season is defined as June 15–October 15, a period that includes all but three of the largest recorded floods. During this season, the average daily rainfall, rainfall intensity, and probability of rainfall are the largest of the year (Fig. 25A–C). Rainfall or occasional snowfall also occurs from early December through early to late March. The pattern, however, is not regular, it lacks repeatability, and precipitation is typically of low intensity. Early April through early June is the driest time of the year, when drought or near-drought conditions prevail. This distinctive annual cycle is repeated in the annual runoff of the San Pedro River, as illustrated in Figure 26. Average discharge, maximum discharge, and sediment load are consistently large during the wet season of June 15–October 15 (Fig. 25A–C).

TABLE 3. LOCATION, ELEVATION, PERIOD OF RECORD, AND PERCENT MISSING DAYS OF EIGHT WEATHER STATIONS AND SAN PEDRO RIVER GAGING STATION USED IN CLIMATE ANALYSIS

Station	Latitude Longitude	Elevation (m)	Period of Record	Missing Days (%)
Apache Powder Co.	31°51' 110°15'	1,125	7/1/1923 2/29/1988	0.90
Benson	31°58' 110°18'	1,095	6/1/1898 5/31/1975	6.04
Bisbee	31°27' 109°55'	1,632	1/1/1895 2/28/1985	25.85
Cochise Stronghold	31°57' 109°57'	1,449	2/1/1899 12/31/1954	34.38
Fairbank	31°43' 110°11'	1,174	7/1/1909 3/31/1973	6.98
Fort Huachuca	31°34' 110°20'	1,423	2/1/1900 12/31/1981	42.56
Patagonia	31°33' 110°45'	1,233	7/1/1921 12/31/1977	2.9
Tombstone	31°42' 110°03'	1,406	2/1/1897 2/29/1988	4.63
San Pedro River at Charleston, Qw	31°37' 110°10'	1,206	3/29/1904 2/29/1988	11.72
San Pedro River at Charleston, Qs	31°37' 110°10'	1,206	7/7/1963 9/30/1975	0

Generally, large floods on the San Pedro River are very much controlled by wet-season rainfall of mid-June to mid-October. The day of the annual flood, the largest flood of the year, is shown in Figure 27. Only 3 of the 73 annual floods shown in the figure occurred after October 15; the remaining 70 floods occurred between earliest July and mid-October. The symbols in Figure 27 show the distribution of floods in three periods, 1916–1930, 1931–1960, and 1961–1989. The 95% confidence interval of the average date of the annual flood for the three periods is July 30–August 29, July 28–August 13, and August 14–September 19, respectively. The confidence intervals of all except the second and third periods overlap, although the difference is not great. Whether this signifies a significant shift to floods later in the season is unknown.

Other workers have noted a change in flood seasonality of the Santa Cruz River (Osborn and Lane, 1984; Roeske and others, 1989; Webb and Betancourt, 1990; Betancourt and Turner, 1993). Seasonality of floods on the Santa Cruz River shifted after 1960 to fall and winter. A similar shift is not evident in the San Pedro River data, although the only three winter floods occurred in the period between 1961 and 1989.

Climatology of wet-season rainfall

Floods of the upper San Pedro River, as discussed above, occur almost exclusively in the wet season June 15–October 15.

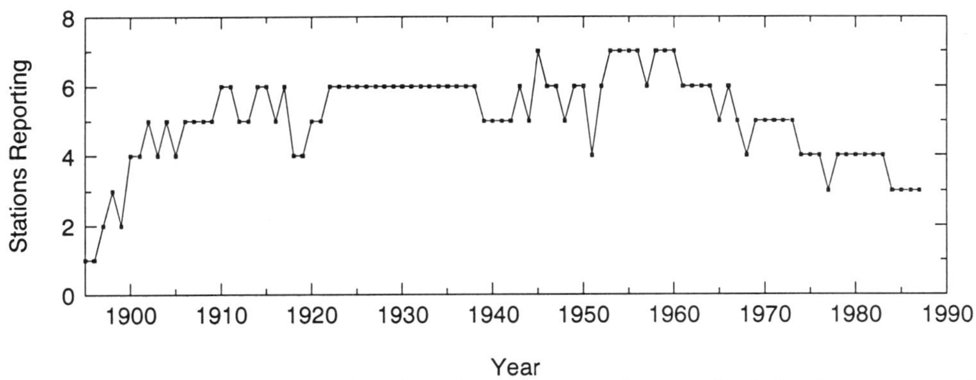

Figure 24. Time series showing number of weather stations reporting annually during the wet season of June 15–October 15, 1895–1987.

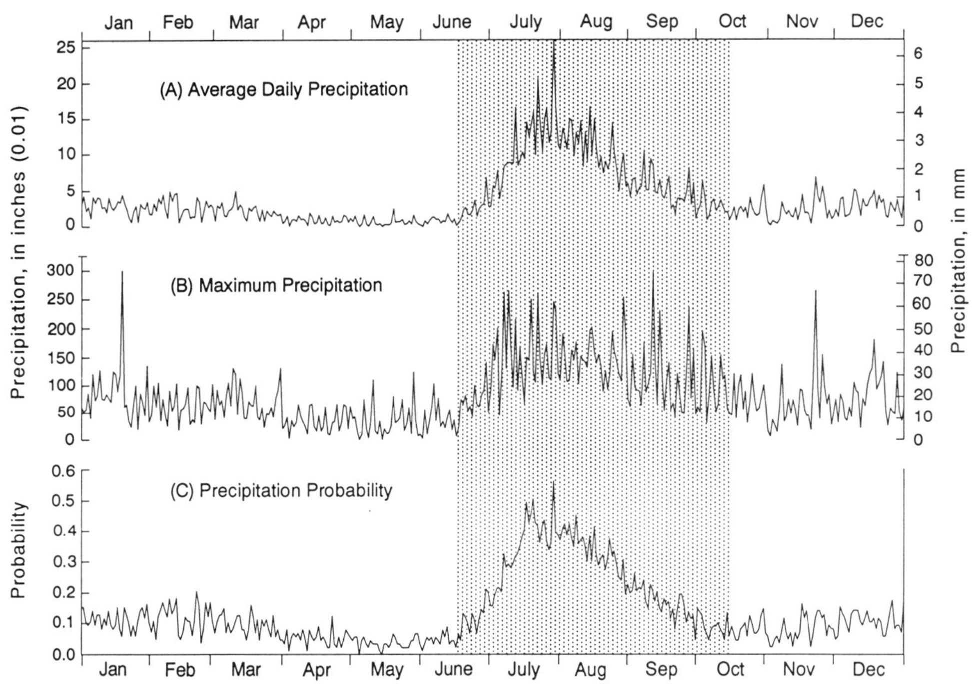

Figure 25. Annual precipitation cycle at Tombstone, based on period of record, 1897–1988. (A) Average daily precipitation greater than a trace (0.01 in, or 0.254 mm). (B) Maximum recorded precipitation for the day. (C) Probability of precipitation, computed from number of days with rain divided by the total number of entries of a particular day. Pattern shows the wet season of June 15–October 15.

According to Webb and Betancourt (1990, p. 10–19), flood-producing rainfall and runoff in southern Arizona result from three large-scale (at times interrelated) atmospheric circulation patterns that typically cause rainfall at different times of the wet season. These patterns are monsoonal circulation, cut-off low-pressure systems, and dissipating tropical cyclones. Flood-producing rain in the early part of the season from July to late August or early September typically results from the Southwest "monsoon," so named because of the similarity with monsoonal rainfall mechanisms elsewhere. Early in the season, the subtropical high-pressure cells shift rapidly northward, advecting moist, tropical air into the Southwest. Strong heating of the moist air produces convectional rainfall that is widespread and locally intense. "Bursts," or an increase of monsoonal rainfall, are related

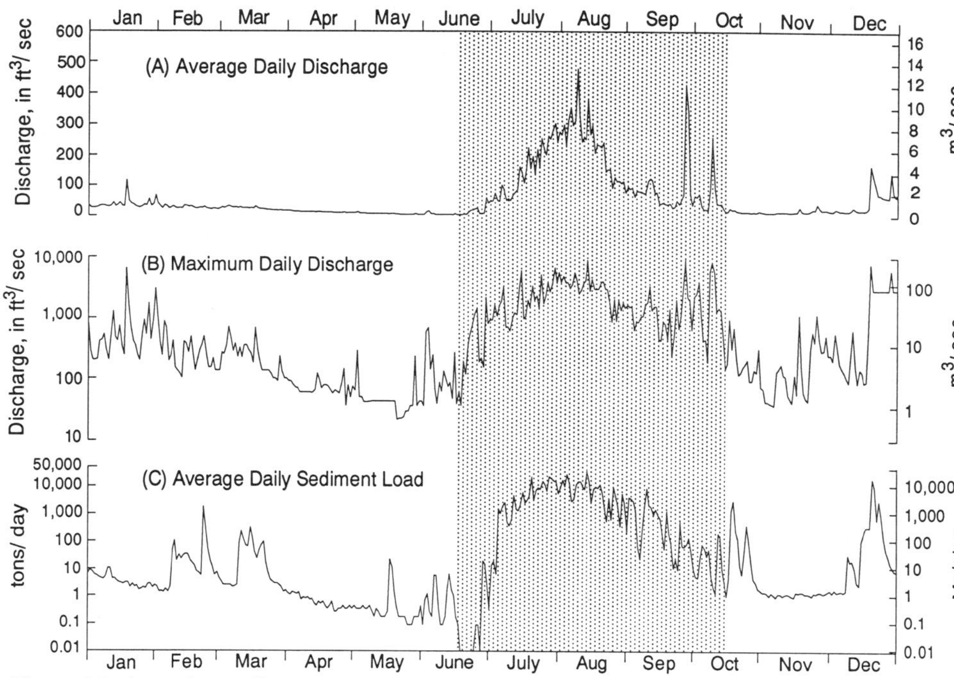

Figure 26. Annual runoff cycle of the San Pedro River at Charleston (Fig. 1), computed from 1916–1988. (A) Average daily discharge. (B) Maximum daily discharge. (C) Average daily sediment load. Pattern shows the wet runoff season of June 15–October 15.

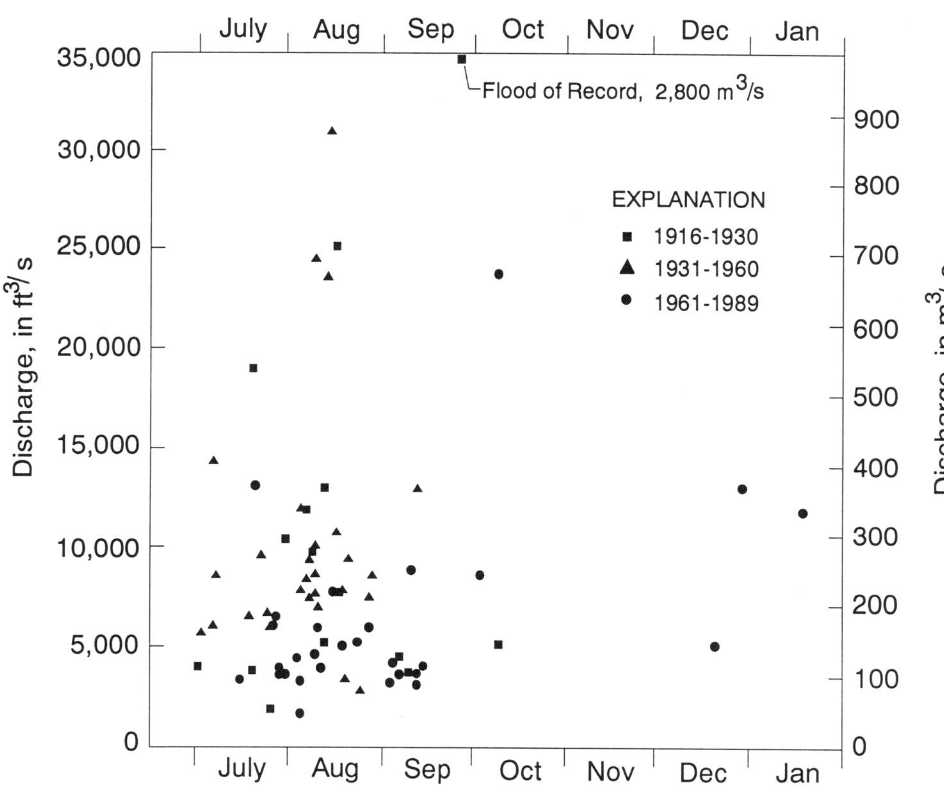

Figure 27. Date of the annual flood of the San Pedro River at Charleston, based on 1916–1987 flood series. All but three floods occurred during the wet season.

to several weak patterns of atmospheric circulation caused by northerly displacement of the Bermuda High or the North Pacific subtropical anticyclone (Carleton, 1986). In contrast, southerly displacement of these high pressure areas cause "breaks," or a suppression of monsoonal rainfall.

In the latter part of the wet season, significant rainfall can result from cut-off low-pressure systems and dissipating tropical cyclones. Cut-off low-pressure systems develop when a high-pressure ridge forms in the eastern Pacific. Low-pressure systems are "cut-off" from the main jet stream and drift south along the west coast until they eventually move inland over Arizona. When they interact with tropical cyclones, conditions are enhanced for extreme precipitation. Tropical cyclones, generated near the equator off the western coast of Mexico, drift west and northwest until they lose energy and dissipate harmlessly over relatively cool water. At times, however, the storms recurve to the north and east, eventually dissipating over Mexico and the southwestern United States. Rainfall is widespread and locally heavy when this happens.

The types of weather systems generating floods on the nearby Santa Cruz River were identified by Webb and Betancourt (1990, Table 8). They found that 53% of the annual floods between 1915 and 1987 were generated by monsoonal rainfall, 24% were related to tropical cyclones, and 23% were generated by rainfall associated with frontal systems. In contrast, 87% of San Pedro River floods resulted from monsoonal rainfall, based on comparison of the San Pedro River flood series with their data. The second largest flood (Fig. 23), having a peak discharge of 880 $m^3\ s^{-1}$ (31,000 $ft^3\ s^{-1}$), on August 13, 1940, resulted from monsoonal rain. Dissipating tropical cyclones produced at least five floods on the San Pedro River. The flood of record (Fig. 23), with a peak discharge of 2,800 $m^3\ s^{-1}$ (98,000 $ft^3\ s^{-1}$), was generated by rainfall originating from a dissipating tropical cyclone. More recently, the largest flood of the post-1955 era (Fig. 23), having a discharge of 670 $m^3\ s^{-1}$ (23,700 $ft^3\ s^{-1}$), also resulted from rainfall associated with a dissipating tropical cyclone. Frontal and cut-off low-pressure systems have produced at least three floods on the San Pedro River. This includes the second largest flood of the post-1955 era (Fig. 23), the flood of December 28, 1984, with a discharge of 370 $m^3\ s^{-1}$ (13,000 $ft^3\ s^{-1}$).

Floods and antecedent rainfall

Rainfall producing the annual flood of the upper San Pedro River is associated with wet spells lasting several days and is regional in extent. Table 4 lists the characteristics of 12-day antecedent rainfall (11 days before and the day of the flood) associated with the annual flood. On average, these floods follow 4 to 6 days of rain that occurs intermittently over the 12-day interval, accumulating between about 36 and 58 mm at the eight weather stations. More than a trace of rain (0.01 in., or 0.254 mm) is reported in the 12-day period in about 90 to 98% of the cases (Table 4). Thus, flood-producing rainfall is not of the local, short-term character usually associated with rainfall in semiarid regions (Graf, 1988, p. 72–73).

TABLE 4. ANTECEDENT RAINFALL STATISTICS. AVERAGE TOTAL RAINFALL, AVERAGE NUMBER OF RAINY DAYS, NUMBER OF YEARS OF RAINFALL DATA, AND PERCENT FLOODS WITHOUT ANTECEDENT RAINFALL AT A STATION*

Station	Average Total Rainfall (mm) ± 95% CI†	Average Days with Rain ± 95% CI†	Years Reporting	Floods Without Rain (%)
Apache Powder Co.	45.6 ± 8.6	5.1 ± 0.6	64	4.69
Benson	36.0 ± 7.4	4.4 ± 2.4	56	7.14
Bisbee	44.9 ± 9.3	5.6 ± 0.8	48	2.08
Cochise	35.8 ± 13.8	4.2 ± 1.1	20	10.0
Fairbank	41.6 ± 10.6	4.3 ± 0.7	53	7.55
Fort Huachuca	50.2 ± 15.3	5.7 ± 0.9	30	3.33
Patagonia	57.7 ± 8.7	6.4 ± 0.8	66	1.52
Tombstone	42.8 ± 8.6	4.7 ± 0.6	69	5.80

*Based on annual flood series of the Charleston gage from 1916–1987. Rainfall accumulated for 11 days before and day of each of 71 floods.

†CI = confidence interval.

A time series of antecedent rainfall and its relation with the annual flood is shown in Figure 28. The long-term pattern of antecedent rainfall (Fig. 28A) suggests that the post-1955 period had somewhat greater rainfall for a given flood. The latter period had two of the largest rainfall totals (1977 and 1983), these were probably greater than the rainfall of 1926, which produced the flood of record. Moreover, the period 1964–1977 was characterized by persistently large antecedent rainfall. This tendency of increased antecedent rainfall after 1955 is inconsistent with the annual flood series (Fig. 23), which has a clear pattern of less frequent large floods after 1955. This conclusion of divergent rainfall and flood patterns is somewhat equivocal, however, because variability in the size of the annual flood is not well explained by antecedent rainfall of the eight weather stations (Fig. 28B). For example, a two-order of magnitude increase of rainfall produces only about a one-order magnitude variation of peak discharge. Additional factors probably influence the size of the annual flood: unrecorded rainfall in the mountains surrounding the basin (Fig. 1), unknown rainfall in Mexico, and variable basin-runoff characteristics.

Seasonal rainfall and runoff

Annual rainfall and runoff were analyzed using three seasons: October 16–February 14 (midfall to late winter), February 15–June 14 (latest winter to spring), and June 15–October 15 (summer to early fall). The seasons were chosen to have equal duration for comparison of precipitation amounts and to emphasize the annual decline of precipitation following the June 15–October 15 maximum. Table 5 lists the average total rainfall

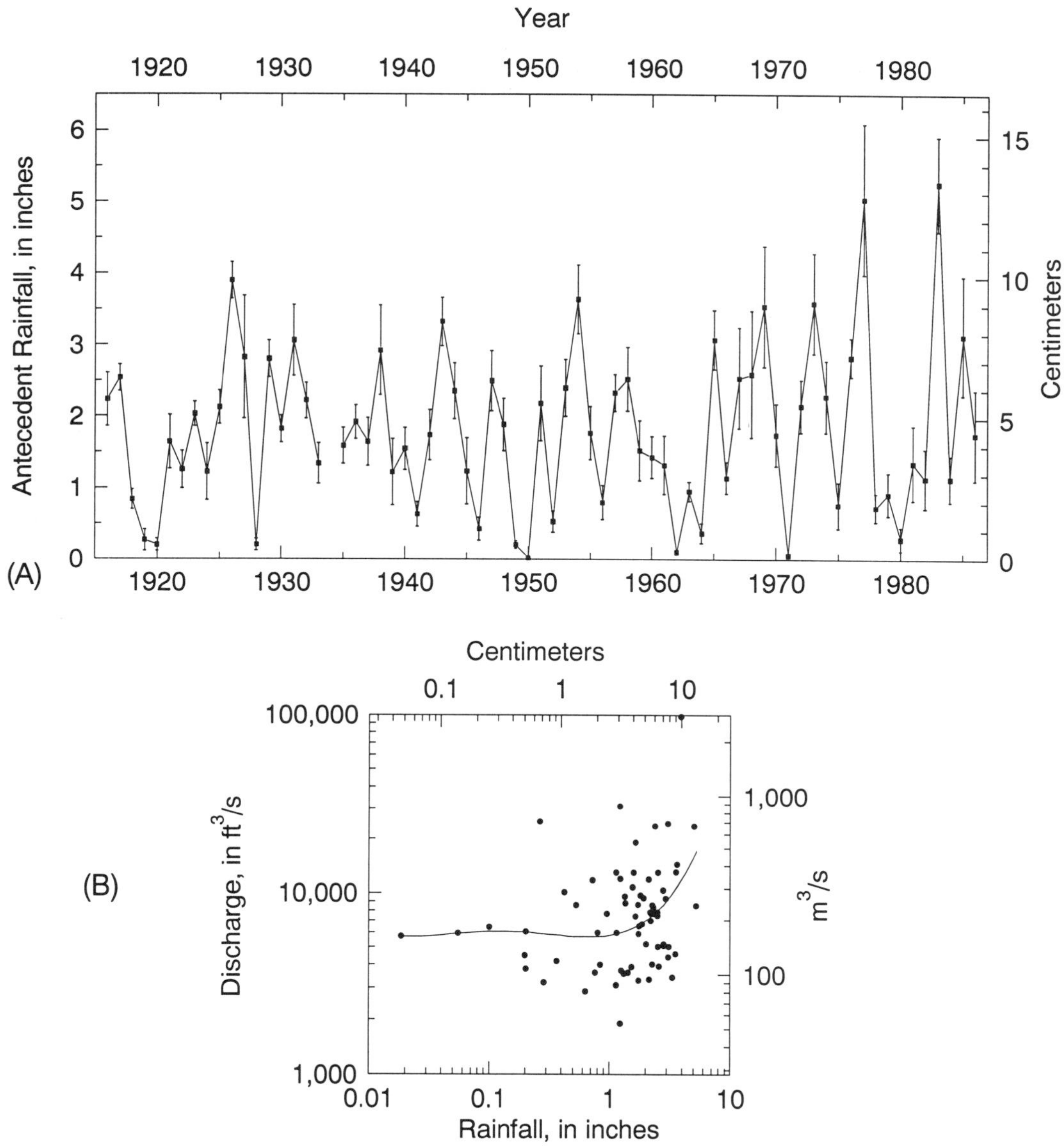

Figure 28. (A) Time series of average-total antecedent rainfall of the eight weather stations (Table 3) during annual flood of San Pedro River, 1916–1986. Vertical bar is the standard error of total rainfall. (B) Annual peak-flood discharge as a function of average-total antecedent rainfall. Variation of peak-flood discharge is not well explained by variation of antecedent rainfall.

of the three seasons at the eight stations. Rainfall during the wet season is on average 5.8 times greater than the late winter to spring dry season of February 15–June 14. Moreover, wet-season rainfall averages 2.8 times greater than rainfall during the moderately moist midfall to late winter season of October 16–February 14.

Time series of the three seasons for each of the eight stations are shown in Figures 29 through 31. The fragmented, broken character of the time series results from missing entries. Nonetheless, comparison of the series shows that they have similar patterns of peaks and troughs, indicating that they are consistent with each other. For regional analysis, the seasonal totals of each reporting station were averaged to form a single annual value.

Figure 32 is a composite of the eight stations that does not have data gaps because when combined at least two stations reported annually (see Fig. 23 for the number of stations reporting annually), which is the minimum necessary to form the average. The midfall to late winter season was characterized by several years of distinctly above average precipitation from 1903–1920 and from 1960–1985 (Fig. 32A). Years with prolonged or extreme drought conditions were 1909–1912, 1916–1917, 1921–1923, 1925, 1934, 1938, 1943–1944, 1951,

TABLE 5. AVERAGE ANNUAL THREE-SEASON RAINFALL OF THE EIGHT WEATHER STATIONS

Station	October 16–February 14 (mm) ± 95% CI*	February 15–June 14 (mm) ± 95% CI*	June 15–October 15 (mm) ± 95% CI*
Apache Powder Co.	71.8 ± 10.4	33.9 ± 6.7	220.1 ± 19.1
Benson	67.5 ± 13.1	30.0 ± 5.7	190.3 ± 15.1
Bisbee	115.2 ± 22.1	55.7 ± 10.3	292.5 ± 20.3
Cochise	74.1 ± 17.7	39.5 ± 13.6	170.6 ± 23.4
Fairbank	57.9 ± 9.5	28.2 ± 6.3	214.3 ± 20.3
Fort Huachuca	101.4 ± 31.5	47.9 ± 14.4	244.1 ± 26.6
Patagonia	112.0 ± 18.1	52.0 ± 7.9	288.6 ± 19.4
Tombstone	82.4 ± 11.6	39.5 ± 7.9	234.0 ± 15.3

*CI = confidence interval.

1954–1956, 1967, and 1970–1971. The latest winter through spring season (Fig. 32B) was dominated by the extremely wet season of 1915, although 1926, 1952, 1958, and 1973 were also unusually wet. Drought conditions prevailed during this season in 1925, 1929, 1933–1937, 1943, 1946, 1949–1950, 1956, 1959–1966, and 1974.

The wet season of June 15–October 15 (Fig. 32C) was dominated by a complex pattern of alternating wet and dry years. Rainfall during the early part of the record from about 1899–1918 was at or below the long-term average. From 1919–1931, rainfall was generally well above normal. After about 1931 until 1953, rainfall oscillated between extremely dry and extremely wet. Unusually dry-year periods were 1947 and 1950–1952. After about 1953, this pattern reversed, with wet years of 1 to 2 yr in duration alternating with moderately dry years. Extremely dry-year periods were 1953, 1973, and 1978–1981. The early period of below normal wet-season rainfall suggests that, on an annual basis, this was a dry period, particularly considering that wet-season rainfall dominates the annual totals (Table 5). However, the deficient wet-season rainfall was balanced by the above average rainfall of the other seasons (Fig. 32A, B).

The average daily discharge of the San Pedro River for the three seasons is illustrated in Figure 33. The two seasons from mid-fall through spring (Fig. 33A, B) had unusually high discharge from 1915–1921, and discharge was intermittently high from 1968–1986. Likewise, wet-season discharge was unusually high in the early and latter parts of the record (Fig. 33C); specifically, 1904, 1914–1921, 1931, 1954–1955, 1958, and 1977. Wet-season discharge decreased since 1960; before 1960, the average daily discharge was 4.4 $m^3 s^{-1}$ (154 $ft^3 s^{-1}$), since 1960, discharge has been only 2.4 $m^3 s^{-1}$ (86.1 $ft^3 s^{-1}$). The difference between the two averages is statistically significant, with $P < 0.0027$.

Beginning about 1951, the short-term pattern of wet-season discharge evidently changed. This is illustrated in Figure 34, which shows the autocorrelation function of streamflow (Fig. 34A) and rainfall (Fig. 34B) for two periods of equal length (1913–1950 and 1951–1987). Before 1951, the typical pattern was a yearly alternation between high and low values, producing a "sawtooth" pattern. In Figure 34A, this pattern is demonstrated by the negative correlation of points 1 yr apart and the positive correlation of points 2 yr apart. After 1950, the typical pattern was a biennial or longer oscillation, as indicated in Figure 34A by the negative correlation of points 2 yr apart and the weak positive correlation of points 6 yr apart. Rainfall has a similar pattern of short-term variation before and after 1951 (Fig. 34B), although the pattern is not as well developed. In short, the duration of below average discharge and rainfall was about 1 yr before 1951, but increased to 2 to 5 yr from 1951 on.

The long-term patterns of rainfall and runoff (Figs. 32, 33) are similar, except for the reduced daily discharge of the wet season (Fig. 33C) after about 1960, a pattern not evident in the wet-season rainfall (Fig. 32C). This reduced daily discharge is quite likely caused by the corresponding decrease of peak-flood discharge, because seasonal discharge is largely a function of the size of the annual flood. For the most part, long-term seasonal rainfall (Fig. 32) is reasonably well correlated with streamflow, particularly the October 16 and June 14 seasons. Figure 35 shows average daily discharge as a function of rainfall for the three seasons. The fall to late winter season (Fig. 35A) is well correlated with discharge; rainfall accounts for about 88% of the variation of discharge during this season. Rainfall accounts for only 32% of the variation of discharge of the latest winter through spring season (Fig. 35B). Wet-season discharge shows considerable scatter; nevertheless, rainfall accounts for nearly 50% of the discharge variation (Fig. 35C), and the long-term runoff pattern (Fig. 33C) effectively reproduces the peaks and troughs of the rainfall time series.

Wet-season rainfall intensity

Wet-season rainfall was analyzed to search for long-term variations of intensity. This type of analysis is used to identify geomorphically significant long-term changes of rainfall not apparent in seasonal or annual data (Leopold, 1951; Cooke and Reeves, 1976; Betancourt and Turner, 1993; Hereford, 1989; Hereford and Webb, 1993). The rationale is similar to statistical analysis of variance in which the rainfall is broken up into components of variation. If seasonal rainfall, for example, is composed of low- and high-intensity rains, then a change in one or both components might enhance certain geomorphic activity without affecting total seasonal rainfall. The geomorphic effect of frequent high-intensity storms is to weaken vegetative cover, increase the likelihood of large floods, and enhance channel widening. On the other hand, fewer intense storms and an increase of

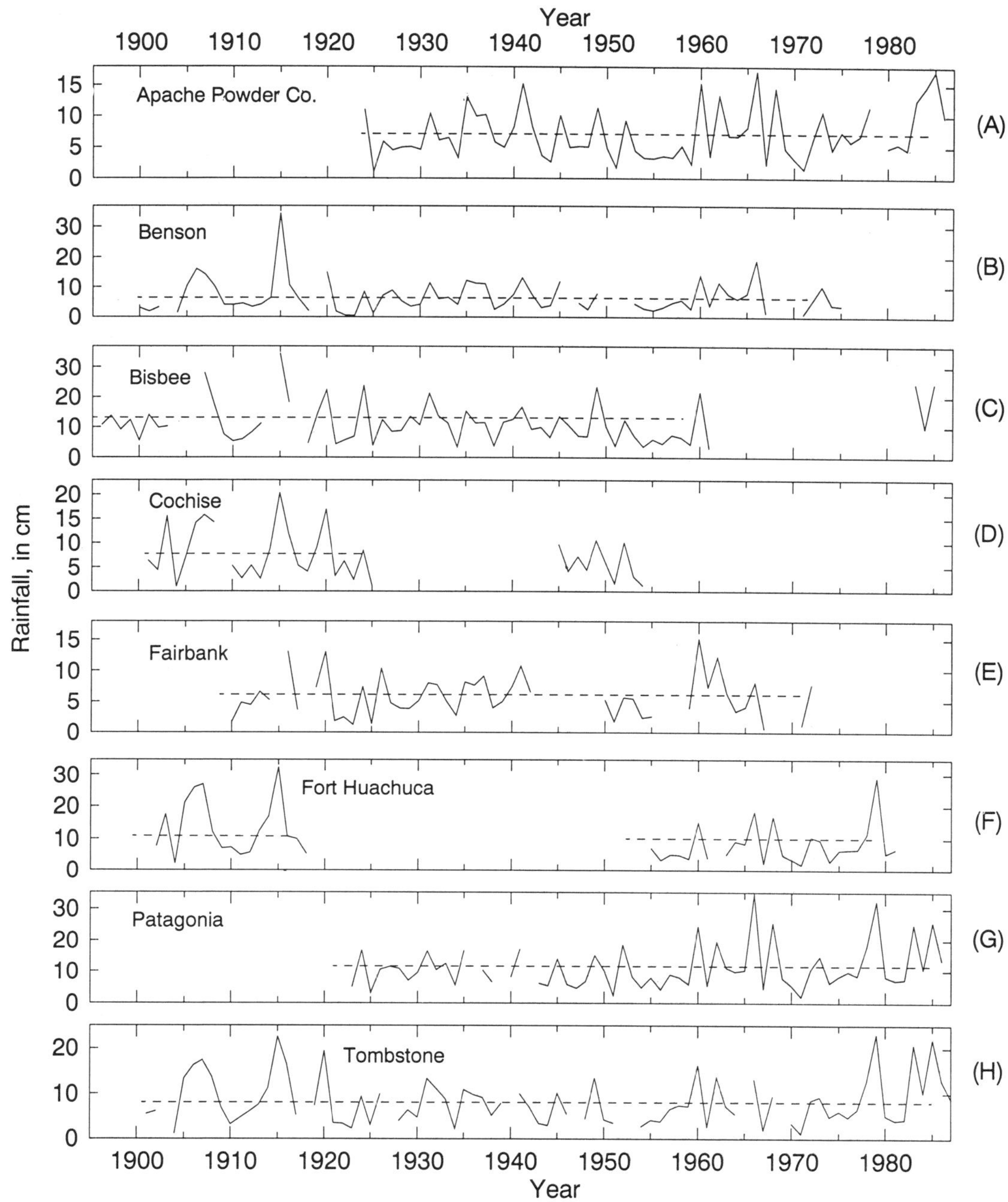

Figure 29. Time series (A through H) of total precipitation for the season October 16–February 14 for the period or record of the eight weather stations. Dashed line is the long-term average. Note correspondence of peaks and troughs among time series (A through H) and lack of any long-term pattern.

low-intensity rainfall may strengthen vegetation, decrease the frequency of large floods, and retard channel widening.

Working with data from only one or two stations, Leopold (1951), Cooke and Reeves (1976), and Betancourt and Turner (1993) defined low-intensity rainfall as the number of days annually with between a trace (0.01 in.; 0.254 mm) and 0.50 in. (1.27 cm); high-intensity rainfall was defined as all days having 1 in. (2.54 cm) or more of rainfall. In a study of 24 stations on the Colorado Plateau, Hereford (1989) and Hereford and Webb (1992) examined the frequency and accumulated wet-season rainfall greater than 0.2 in. (0.5 cm).

Leopold (1951), using a long precipitation record from Santa Fe, New Mexico, found that, whereas long-term annual precipitation did not change, rainfall intensity varied over the

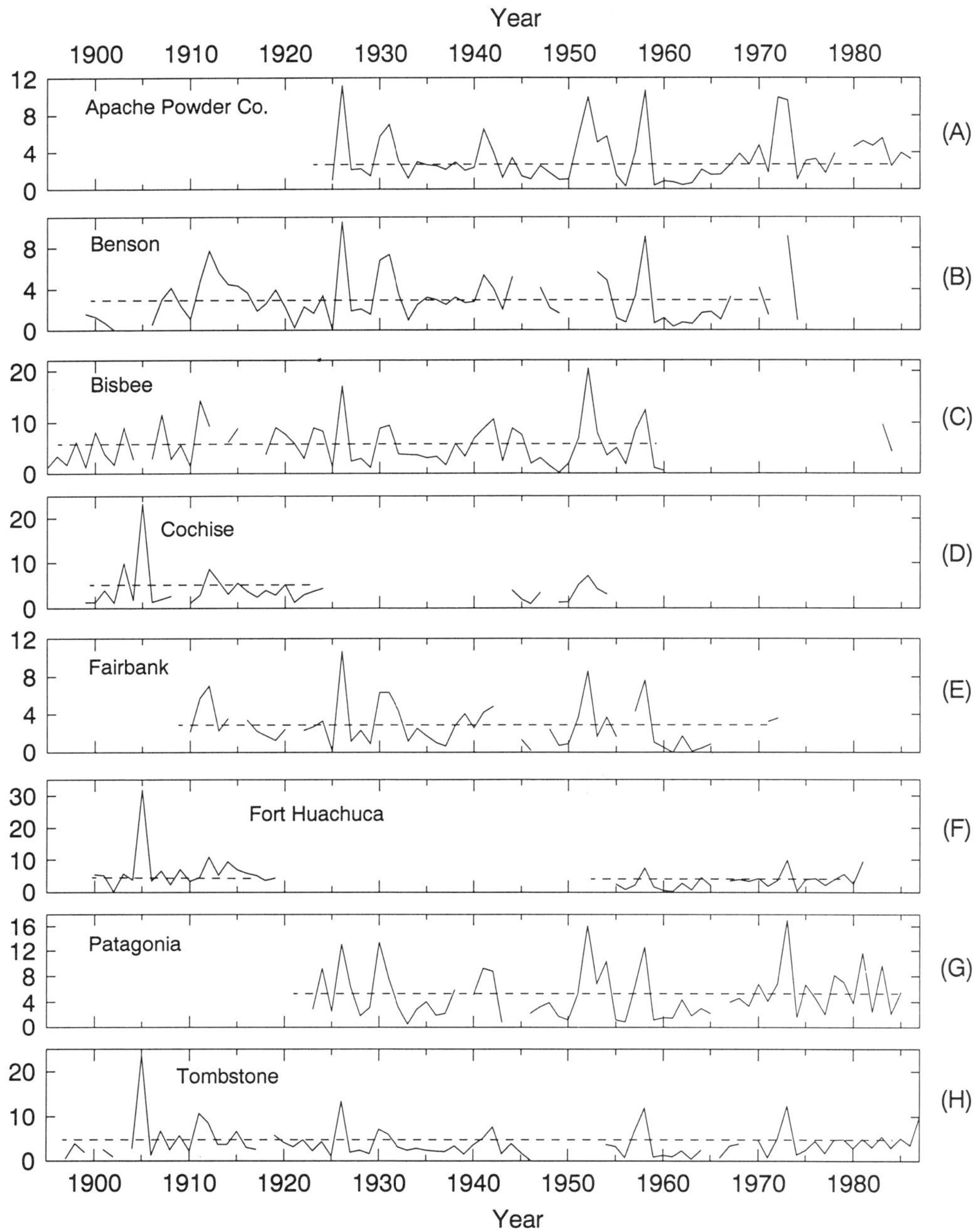

Figure 30. Time series (A through H) of total precipitation for the season February 15–June 14 for the period of record of the eight weather stations. Dashed line is the long-term average. Note correspondence of peaks and troughs among the time series (A through H) and lack of any long-term pattern.

period of record. Precipitation during the latter part of the 19th century was dominated by high-intensity events at the expense of low-intensity precipitation. Cooke and Reeves (1976) and Betancourt and Turner (1993) obtained similar results from a long precipitation series in Tucson. These workers concluded that high-intensity rainfall during the end of the 19th century was an important factor in stream entrenchment. In an analysis of wet-season rainfall of the Colorado Plateau, Hereford (1989) and Hereford and Webb (1992) found that rainfall frequency decreased in the 1930s through early 1940s. These changes were coincident with a decrease of stream discharge, peak flow, and sediment load of Colorado Plateau streams.

Categories of rainfall intensity reflect the individual workers' notion of geomorphically or hydrologically significant rainfall.

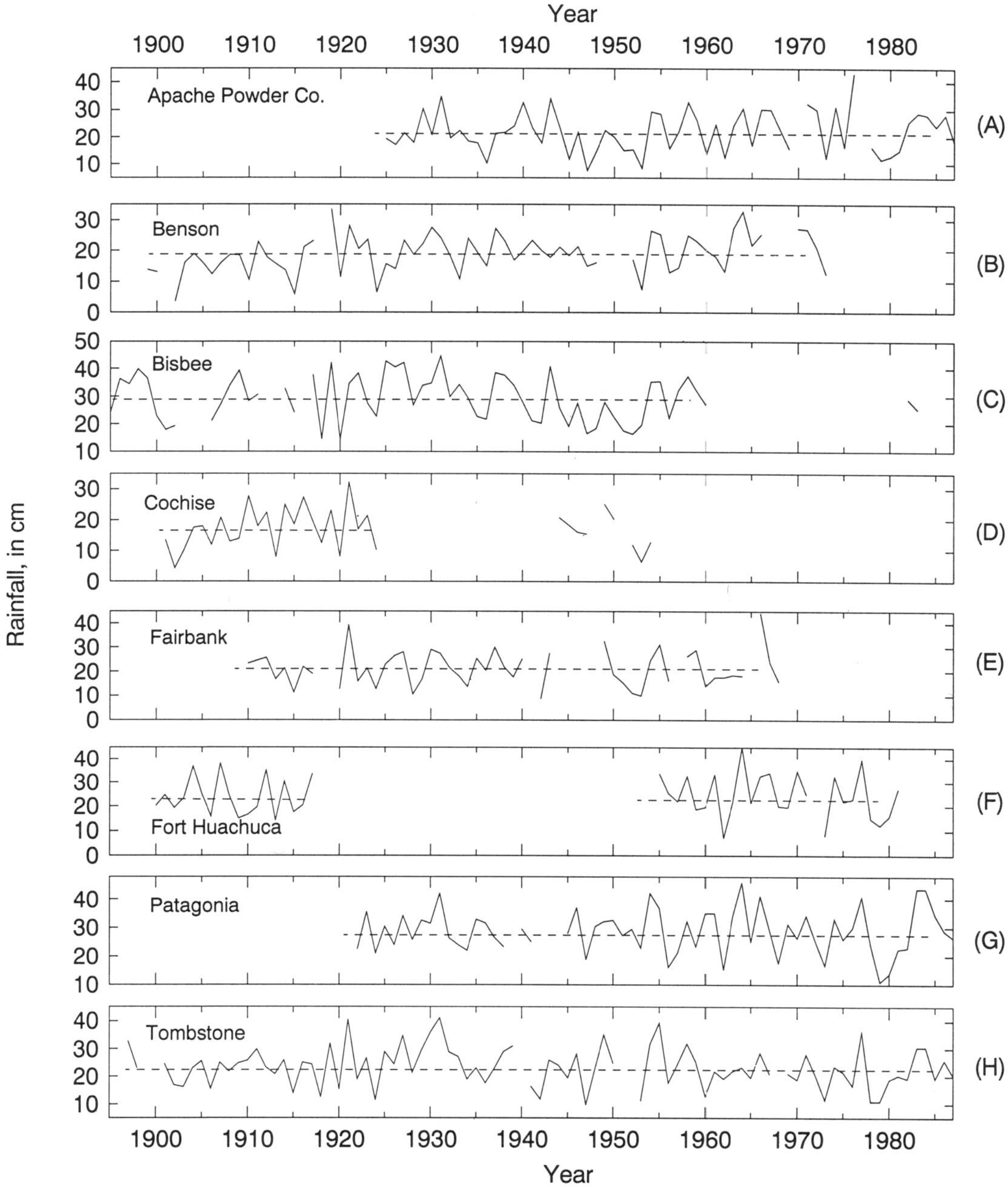

Figure 31. Time series (A through H) of total rainfall for the season June 15–October 15 for the period of record of the eight weather stations. Dashed line is the long-term average. Note correspondence of peaks and troughs among the time series (A through H) and lack of any long-term pattern.

Generally, the categories do not have a physical or empirical basis; rather, they merely represent components of rainfall variation that might have geomorphic or hydrologic significance. Inferences drawn from biologic evidence support the contention that low-intensity rainfall is probably the most important category for enhancing plant growth. A study of desert grasslands of the southern Colorado Plateau, an area having somewhat warmer average summer temperatures than the upper San Pedro River valley, found that moisture was the primary factor limiting grass productivity (Davey, 1980). Moreover, a soil moisture threshold must be exceeded for at least 3 days for a summer growth pulse to occur. Because of the relatively high number of days annually having low-intensity rainfall, this category is most likely to exceed the crucial 3-day threshold.

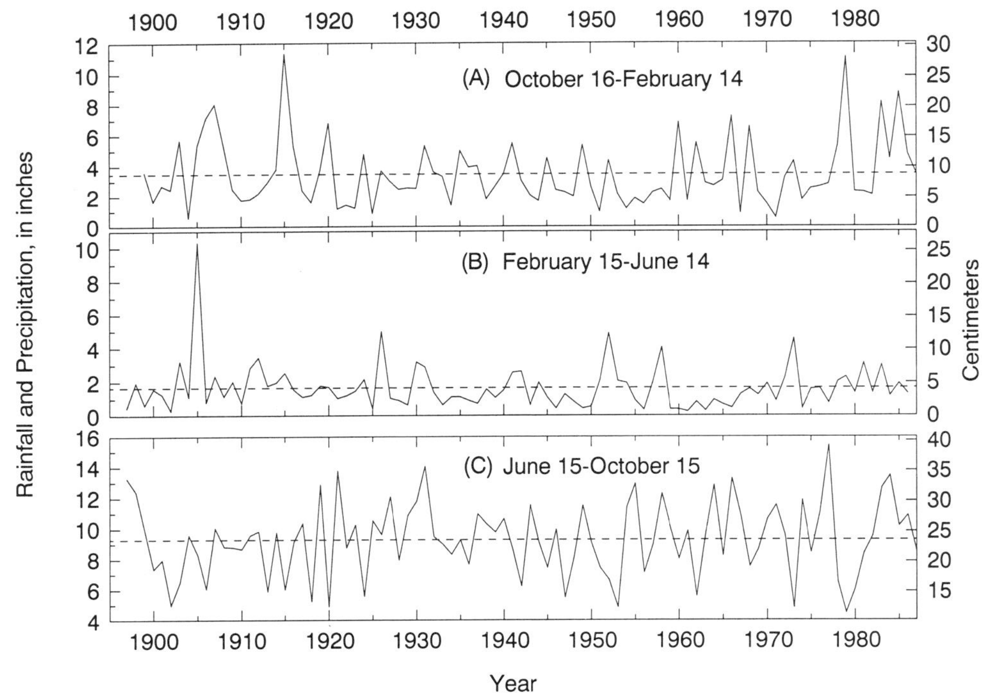

Figure 32. Time series (A through C) of average precipitation and rainfall by season computed by averaging the rainfall by season and station. Dashed line is the long-term average.

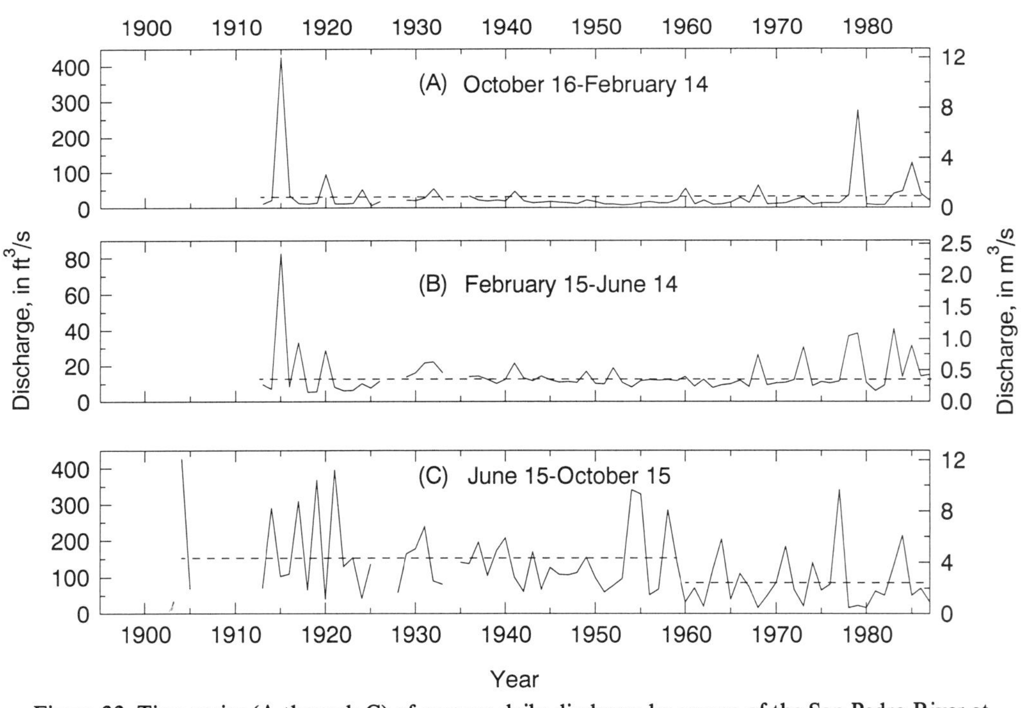

Figure 33. Time series (A through C) of average daily discharge by season of the San Pedro River at Charleston. Note the decrease of wet-season discharge (C) beginning in 1960. Dashed line is the long-term average.

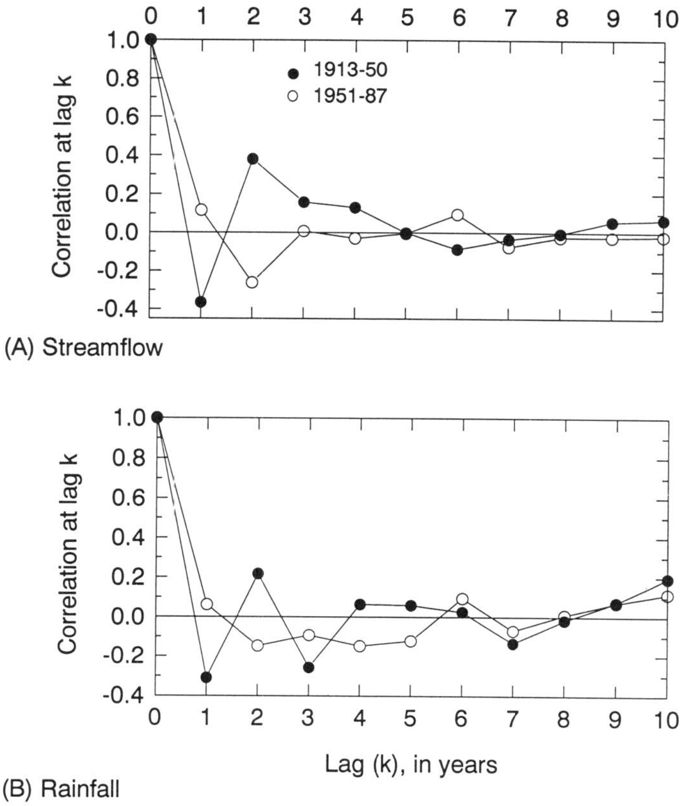

Figure 34. Autocorrelation function of wet-season streamflow (A) and rainfall (B) for two periods of equal length centered at 1951. Correlation is the Spearman correlation coefficient of points k years apart.

Studies show that the frequency of low-intensity rainfall and the long-term pattern of seasonal rainfall are probably linked to germination and survival of rangeland grasses. In southeast Arizona, germination and emergence of grasses typically follow single storms or groups of closely spaced storms with 20 mm or more of rain (Cox and Jordan, 1983). During this 10-yr experiment, wet-season rainfall alternated annually between wet and dry. This pattern resulted in a 84 to 90% decrease in forage production and a 17 to 28% decrease in plant density. In addition, the frequency of rainfall is critical for viability of germinated seedlings. After germination most grasses have difficulty surviving dry spells longer than 7 days (Frasier and Woolhiser, 1990). The possibility of long dry spells is reduced by the frequent occurrence of low-intensity rainfall.

In this study, rainfall intensity is defined as the number of days of rain within a specified size range at a given station. Three rainfall intensity categories were defined: low, intermediate, and high. These correspond to the nonoverlapping 80th, 80–95th, and 95–100th percentiles of the cumulative-distribution function of daily wet-season rainfall. Table 6 lists the size range of the three intensity categories for the period of record of the eight stations.

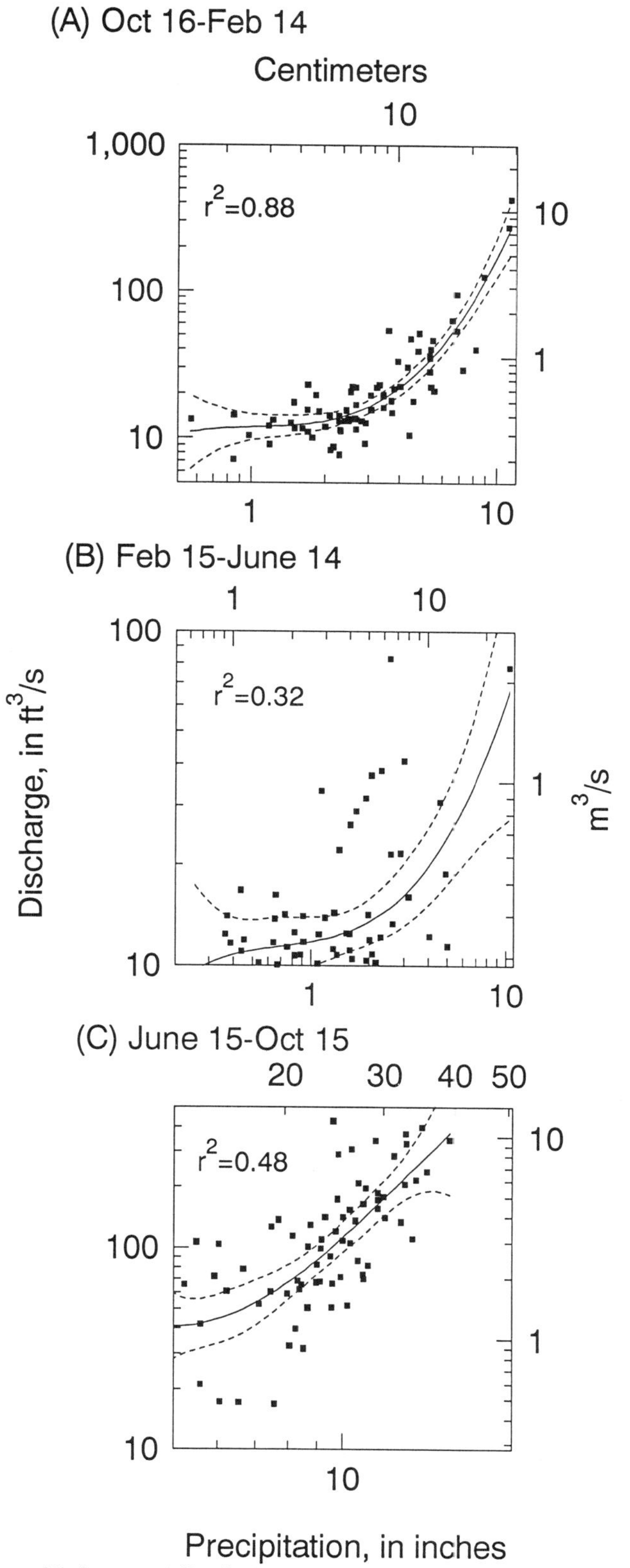

Figure 35. Average daily discharge as a function of rainfall of the three seasons, (A) through (C). Dashed lines show 95% confidence interval of the regression (solid line).

TABLE 6. DAILY WET-SEASON RAINFALL OF THE THREE INTENSITY CATEGORIES FOR RECORD PERIOD OF THE EIGHT WEATHER STATIONS

	Intensity Category		
	Low	Medium	High
	All Daily Wet-season Rainfall		
Station	0 < 80%* (in.) *(mm)*	80 < 95% (in.) *(mm)*	≥95% (in.) *(mm)*
Apache Powder Co.	0.47 *11.94*	1.00 *25.40*	1.00 *25.40*
Benson	0.50 *12.7*	1.00 *25.40*	1.00 *25.40*
Bisbee	0.51 *12.95*	1.21 *30.73*	1.21 *30.73*
Cochise	0.50 *12.7*	1.08 *27.43*	1.08 *27.43*
Fairbank	0.50 *12.7*	1.05 *26.67*	1.05 *26.67*
Fort Huachuca	0.49 *12.45*	1.10 *27.94*	1.10 *27.94*
Patagonia	0.42 *10.67*	0.94 *23.88*	0.94 *23.88*
Tumbstone	0.48 *12.19*	1.05 *25.67*	1.05 *25.67*

*Trace of rainfall (0.01 in.) to indicated value.

The advantage of using percentiles is that variation of intensity among the stations is equalized; thus, each station contributes the same count. The size categories vary slightly among the stations; nevertheless, the size limits of the low- and high-intensity categories are close to those (0.5 and 1.0 in., respectively) thought to be geomorphically significant. For each station and each wet season, rainfall was accumulated and the days of rainfall of the three intensities were counted. The seasonal total rainfall and frequency were averaged to form a single value of total rainfall and intensity for each season.

Rainfall frequency of the three intensity categories is shown in Figure 36, and the relation between frequency and accumulated rainfall of the three categories is shown in Figure 37. The latter figure shows that total rainfall is directly and closely related to frequency in all intensity categories. The long-term average of low-intensity rainfall (Fig. 36A) is about 23 days yr^{-1}. From about 1897–1928, this intensity category was below to only slightly above average. From about 1929–53, low-intensity rainfall gradually declined from above to below average. Beginning in 1957, rainfall shifted to consistently above average until about 1967. Thereafter, low-intensity rainfall oscillated above and below average for periods of 2 to 5 yr.

The long-term variations of low-intensity rainfall in Figure 36A are possibly significant in terms of grass productivity, which in turn might affect runoff. A reduced number of low-intensity storms seasonally would lessen the possibility of reaching and passing the soil moisture threshold necessary for a growth pulse. A high frequency of low-intensity storms would have the opposite effect, increased possibility of growth spurts, which would improve the grass cover.

The long-term average of intermediate rainfall intensity is about 5 days yr^{-1} (Fig. 36B). This rainfall category has a complex pattern of annual variation. From 1900–1928, intermediate rainfall intensity oscillated annually between above and below average. Beginning in 1929, this pattern was replaced by one having a 2- to 3-yr oscillation pattern. High-intensity rainfall on average occurs only 1 to 2 days yr^{-1} (Fig. 36C). This rainfall also has a complex pattern of annual variation that lacks an identifiable long-term trend. Several years, however, were characterized by an unusual number of high-intensity events: 1896, 1921, 1955, and 1977.

Wet-season runoff of the San Pedro River at Charleston (Fig. 1) was analyzed in a manner similar to rainfall. The number of days of seasonal runoff in the nonoverlapping 80th, 80–95th, and 95–100th percentiles were counted. In a manner similar to rainfall intensity, these discharge patterns were classified as small (<101 ft^3 s^{-1}, <2.86 m^3 s^{-1}), intermediate (101 <578 ft^3 s^{-1}; 2.86 <16.4 m^3 s^{-1}), and large (≥ 578 ft^3 s^{-1}; ≥ 14.4 m^3 s^{-1}), respectively. Figure 38 shows the long-term seasonal variation of the three runoff categories. Small discharge events occur on average about 66 days yr^{-1} (Fig. 38A); intermediate, about 18 days yr^{-1} (Fig. 38B); and large events, about 6 days yr^{-1} (Fig. 38C). The annual sequence of large events reproduces closely average daily discharge (Fig. 33C), indicating that seasonal discharge is mainly the result of a relatively few large events. The number of large events decreased substantially from seven to eight events yr^{-1} before 1960 to three to five events yr^{-1} after 1960. The difference between the means of the two periods is significant with $P < 0.0008$, a result similar to the decrease of average discharge since 1960. This decrease of large runoff was matched by a gradual decrease of intermediate events beginning between about 1955.

Figure 39 illustrates the temporal variation of standardized rainfall intensity and runoff. Separation of runoff events by intensity or duration is not as clear as separation of discreet rainfall events because of the delay between rainfall and runoff. Nevertheless, high-intensity rainfall typically precedes large runoff by only several days (Table 4) and is clearly associated with the resulting runoff. Over the period of record, high-intensity rainfall is moderately well correlated with the frequency of large runoff (Fig. 39), and rainfall variation accounts for 43% of runoff variation. Although the correlation between the two is only modest, the annual runoff pattern is similar to the pattern of high-intensity rainfall (Fig. 39). However, the reduced duration of large runoff beginning in 1960 is not evident in the pattern of high-intensity rainfall (Fig. 35C, 39). This lack of correspondence is probably the result of changing channel and basin conditions.

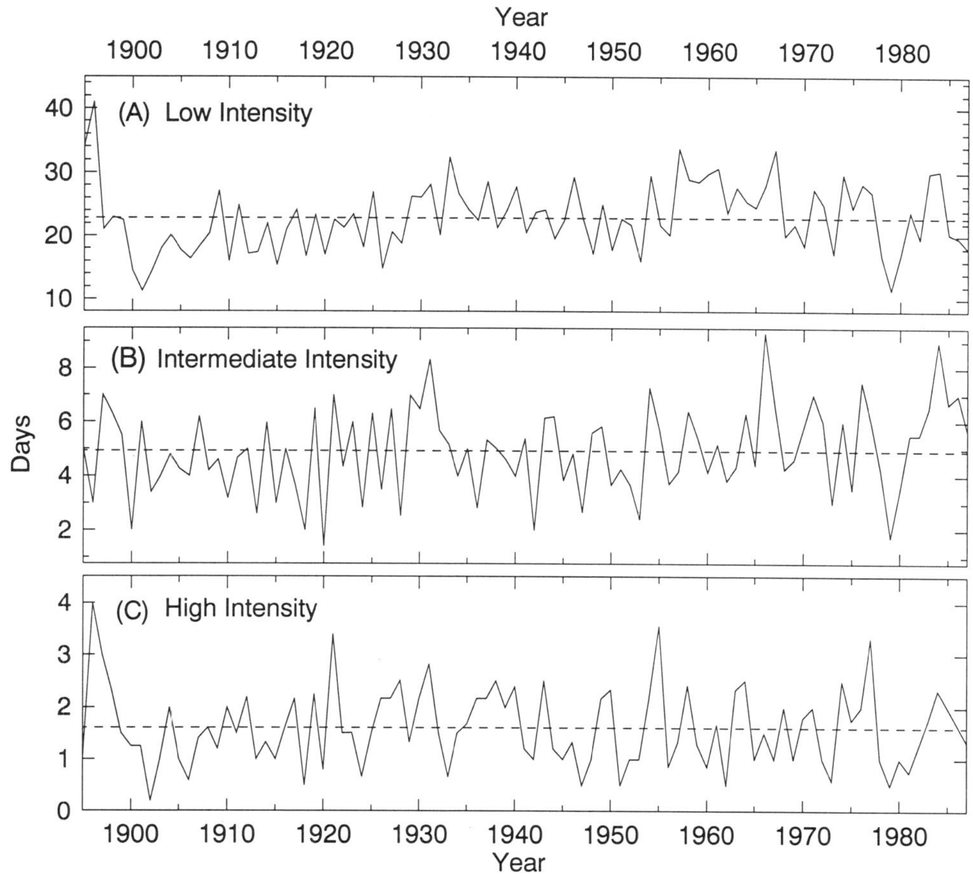

Figure 36. Time series of wet-season rainfall intensity. (A) Days annually having low-intensity rainfall. (B) Intermediate-intensity rainfall. (C) High-intensity rainfall. Dashed line is long-term average.

The changed response of runoff duration to rainfall frequency is evident in Figure 39. For about the first two-thirds of the record, a small increase of rainfall produced a relatively large increase of runoff duration. This effect is well developed in 1913, 1917, and 1919, when rainfall less than one standard deviation above average produced runoff whose duration ranged from 1.5 to 2.5 standard deviations above average. In the latter one-third, however, a large increase of rainfall produced a relatively small increase of runoff duration, an effect well illustrated in 1977, when rainfall more than two standard deviations above average produced runoff of only average duration. This changed response of runoff to rainfall is illustrated in Figure 40, a time series of the cumulative difference of standardized rainfall and runoff. The response of runoff duration to rainfall frequency began to reverse about 1959–1960, a persistent reversal unlike that of 1935–1940 (Fig. 40).

The relation of rainfall to runoff plotted by period, 1913–1960 and 1961–1987, is shown in Figure 41. The rainfall intensity 95% confidence intervals overlap; however, runoff frequency confidence intervals do not, indicating that frequency was greater during the early period. Rainfall intensity did not change between the two periods, but the average runoff duration decreased by as much as 5 days in the 1961–1987 period.

Summary of climate and runoff

Rainfall during the wet-season of June 15–October 15 is the most important of the year in terms of flood generation and geomorphic processes in the upper San Pedro River. During this season rainfall and runoff are the largest of the year, and 96% of the annual floods have occurred during this season (Fig. 27). The annual flood is associated with wet spells that are regional in extent. These wet spells consist of 4 to 6 days of intermittent rainfall (Table 4) and typically occur during a "burst" of monsoonal activity. Rainfall antecedent to the annual flood has probably increased, with the largest accumulated antecedent rainfall occurring in the post-1955 era (Fig. 28A). The effect of these changes on the size of the annual flood is not clear because of the large variability of flood size relative to accumulated antecedent rainfall (Fig. 28B). Nonetheless, the size of the annual flood was substantially smaller after 1955, even though antecedent rainfall of the two largest floods of the post-1955 era was as large or larger than any antecedent rainfall of the pre-1955 era.

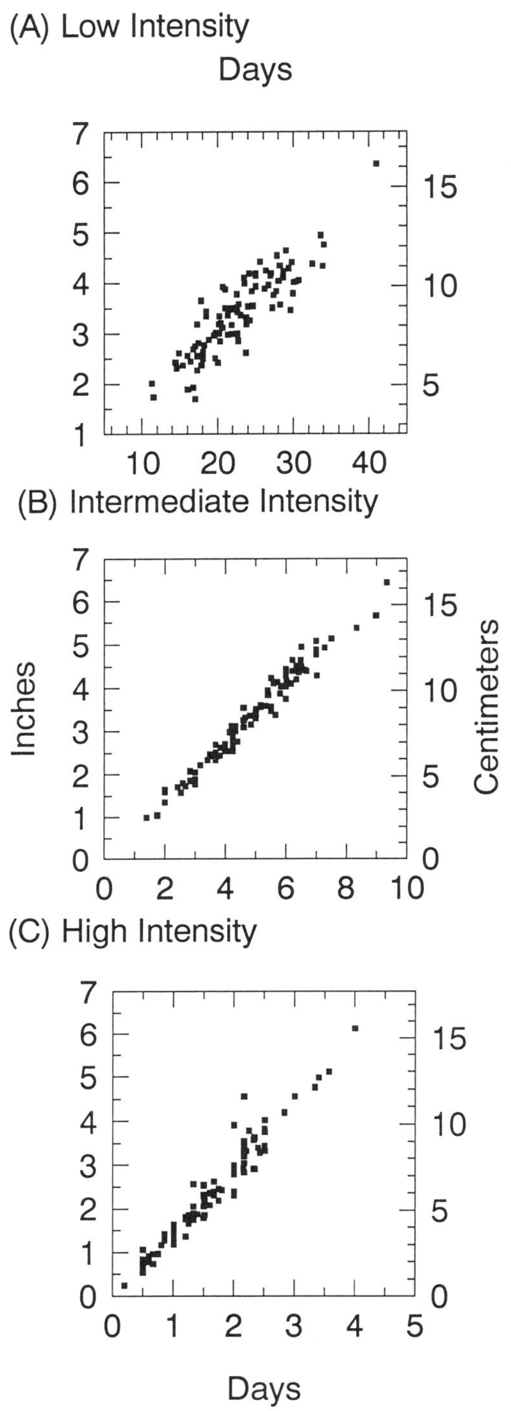

Figure 37. Accumulated rainfall as a function of rainfall intensity, by intensity category. (A) Accumulated low-intensity rainfall. (B) Accumulated intermediate-intensity rainfall. (C) Accumulated high-intensity rainfall.

Analysis of seasonal precipitation did not reveal any long-term pattern of variation for any season for any of eight weather stations (Figs. 28 through 31). Moreover, a combined record without missing seasonal entries confirms the conclusion of no detectable long-term change for the period 1897–1987 (Fig. 32). The seasonal discharge of the San Pedro River also lacks any long-term pattern of variation (Fig. 33), except for the wet season, which shows a decrease of average daily discharge beginning in 1960. This decrease is approximately coincident with the decrease of peak-flood discharge.

Since about 1951, the short-term pattern of wet-season rainfall and runoff has probably changed (Fig. 34). Before 1951, the pattern of rainfall and runoff was mainly annual alternation of wet and dry years, imparting a "saw-tooth" pattern to the respective time series. Since 1951, this short-term pattern was replaced by one of wet and dry years alternating biennially or somewhat longer, producing relatively extended periods of above or below average conditions. The early pattern of alternating wet and dry years was probably detrimental to the growth of range grasses, which in turn could have increased basin runoff. As previously mentioned, this pattern causes a substantial decline in grass density and production (Cox and Jordan, 1983).

Wet-season rainfall intensity (Table 6) generally lacks any long-term pattern of change (Fig. 36), except for low-intensity rainfall. The number of low-intensity events annually was generally below the long-term average during the early part of the record from the period 1897–1928. In the latter part of the record, from 1954–1967, this category was above the long-term average. Likewise, wet-season runoff intensity lacks a long-term pattern of change (Fig. 38), except that large runoff events decreased in number after about 1960.

Finally, the relation between wet-season rainfall and runoff intensity has probably changed since 1960 (Fig. 40). Before 1960, 1 to 2 days of high-intensity rainfall would produce 7 to 8 days of large runoff (Fig. 41). Since 1960, however, the same rainfall duration produces on average only about 3 to 4 days of large runoff. Thus, runoff duration has declined without a corresponding change of rainfall duration. If any change of rainfall has occurred, it is toward greater rainfall preceding the annual flood. In the post-1955 era, rainfall may have increased, or at the very least remained constant, while the size of peak floods (Fig. 23) and average daily discharge actually decreased (Fig. 33B).

Climate, entrenchment, and channel widening

The climatic factors related to initial entrenchment and the beginning of channel widening between 1890 and 1908 are unknown. Systematic collection of weather data in the study area began only in 1895, and climate of the pre-entrenchment and postentrenchment era cannot be compared. Although the record analyzed here is essentially the climate of the postentrenchment era, the climate record is broadly similar to others in the Southwest that span the period of widespread entrenchment and arroyo cutting.

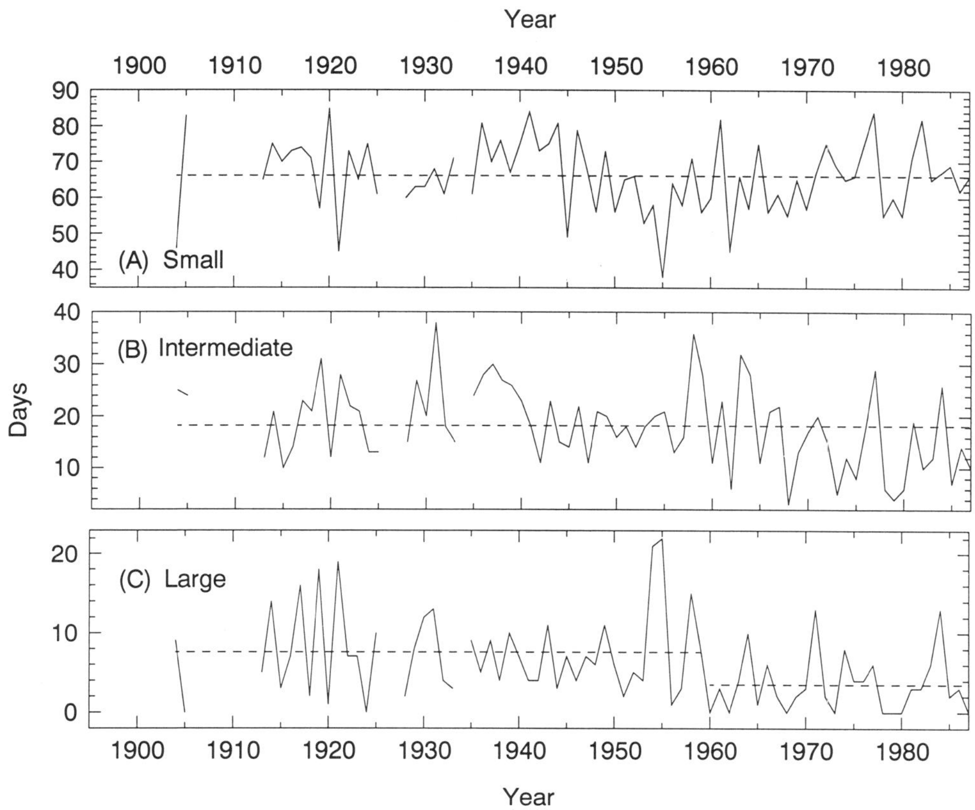

Figure 38. Time series of wet-season runoff frequency classified by percentiles as with rainfall intensity. (A) Days annually having small discharge. (B) intermediate discharge. (C) Large discharge. Note decline of large discharge (C) beginning in 1960 and similarity with high-intensity rainfall (Fig. 36C).

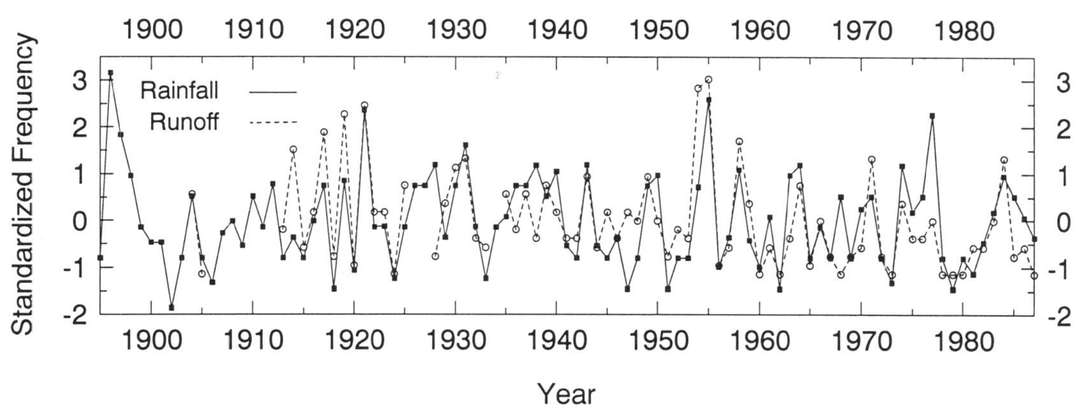

Figure 39. Time series of standardized rainfall and runoff frequency showing the relatively close correspondence between rainfall intensity and runoff.

As previously discussed, rainfall at Tucson (Cooke and Reeves, 1976; Betancourt and Turner, 1993) and Santa Fe, New Mexico (Leopold, 1951), was characterized by relatively few low-intensity events at the time of initial entrenchment through the early part of the 20th century. This is similar to the upper San Pedro River valley where wet-season rainfall intensity was below the long-term average until about 1930. In addition, the annual alternation of wet and dry years that was typical until about 1951 might have adversely affected vegetation by failure during a dry year of young plants established during the previous wet year. This effect was probably exacerbated by grazing practices that favor adding cattle during a wet year and discourage removing them during a dry year (Osborn and Lane, 1984). The annual alternation of wet and dry years, along with grazing pressure,

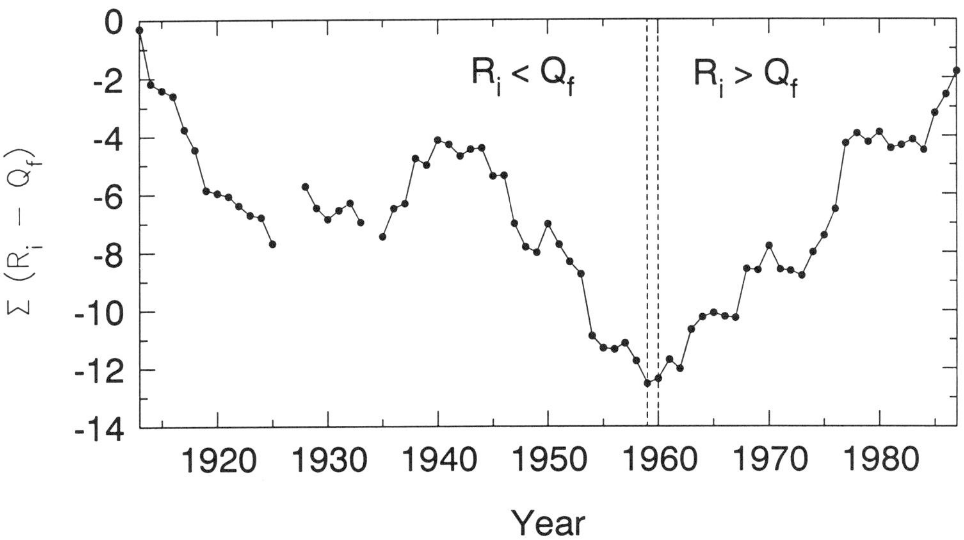

Figure 40. Cumulative departure of rainfall intensity (R_i) and runoff frequency (Q_f) showing the reversal of runoff response in period 1959–1960.

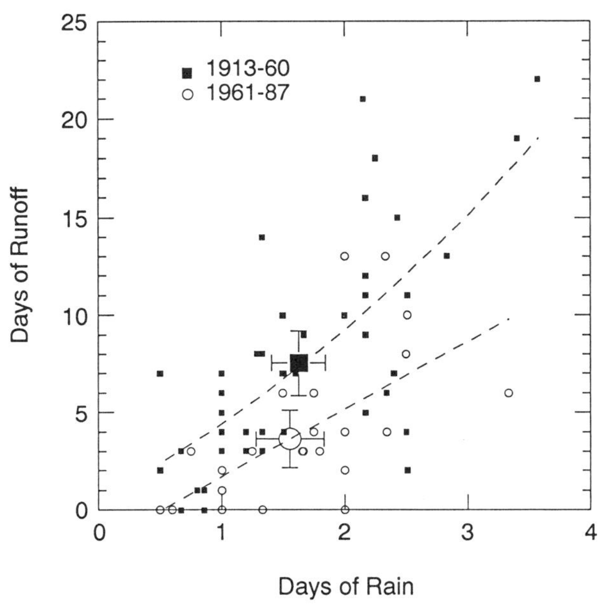

Figure 41. Frequency of high runoff as a function of high-intensity rainfall for two periods, 1913–1960 (solid squares) and 1961–1987 (open circles). Dashed lines are a second- and first-order regression of the 1913–1960 and 1961–1987 periods, respectively. Large square and open circle with bars are the average and 95% confidence intervals, respectively.

probably compounded the effect of reduced low-intensity rainfall, thereby increasing runoff and erosion through failure of the vegetative cover to develop adequately. Finally, high-intensity rainfall in the valley was not above normal during the early part of this century; however, rainfall during the midfall to late winter was unusually high, resulting in large, unrecorded floods out of the wet season. These floods are unrecorded, as they occurred before tabulation of annual floods.

The decline of channel widening and decreased frequency of large floods since 1955 suggest that entrenchment and widening have ended, but it seems unlikely that these changes were directly related to climate as measured by long-term variation of seasonal rainfall. Nevertheless, subtle changes of rainfall intensity and the short-term annual variation of rainfall could have affected growth of in- and extra-channel vegetation, which in turn could have reduced flood frequency. Low-intensity rainfall was consistently above average for the decade 1957–1967. In addition, after about 1951, the short-term pattern of rainfall was for 2 to 5 years of above or below average conditions, which might enhance vegetation by allowing it to become firmly established during a run of wet years. These factors give vegetation the edge over the destructive effects of floods both in and outside of the channel. Increased or improved vegetation cover would reduce basin runoff, and less water would be available to spread over a gradually widening, increasingly vegetated channel. The combined effect of these factors is evident in the reduced frequency of large floods after 1955 and in the reduced duration of large runoff after 1960.

CHANNEL WIDENING AND EQUILIBRIUM

Widening of the San Pedro River channel could not continue indefinitely. Once the channel cross section is capable of transporting the water and sediment load of the postentrenchment discharge regime, it should stabilize and cease to widen significantly. The negligible rate of channel enlargement since about 1955 (discussed in a previous section) indicates that the widening process has ended or slowed greatly. In terms of geomorphic equilibrium, the river system has adjusted to the entrenchment disturbance and has probably attained, or is close to attaining, a new equilibrium with a quasi-stable channel configuration.

This transition from pre- to postentrenchment equilibrium is analyzed diagrammatically in Figure 42, which is based on Graf (1977) and Knighton (1984, p. 179). The morphology of the channel is controlled largely by the frequency of channel-forming floods (the control variable). The effect of an increase of the control variable is to increase the response variable (channel area) after a reaction or lag time. Thus, the pre-entrenchment equilibrium was disturbed by a change of flood frequency probably beginning in the early 1880s, when destructive floods were first noted in the upper San Pedro River valley. An additional disturbance with unknown effect was the 1887 earthquake in northern Sonora, Mexico. The reaction time to these disturbances began about 1880 and lasted until entrenchment began between 1890 and 1908. The period of disequilibrium (Fig. 42) and rapid increase of channel area (Fig. 22) is the relaxation time, or the time to attain an approximate equilibrium. The relaxation time was about 55 yr, assuming that entrenchment began by 1900 and that the channel was essentially stabilized by 1955. Finally, the increase of channel area is approximately an exponential function of time, as suggested by Figure 22. The increase, therefore, follows a "rate law," which describes the time-dependent adjustment of many disturbed physical systems (Graf, 1988).

The relaxation time for channel stabilization was probably controlled by factors influencing the frequency of channel-forming floods. Flood frequency is affected by climate, land use, and feedback between vegetation and channel. As the channel expands, more room is provided for growth and establishment of riparian vegetation, which reduces peak-flood discharge (Burkham, 1976). In addition, larger channel area increases transmission losses, compounding the influence of vegetation. This feedback process shortens the time to stabilization, because vegetation increases boundary shear stress, eventually minimizing further bank erosion.

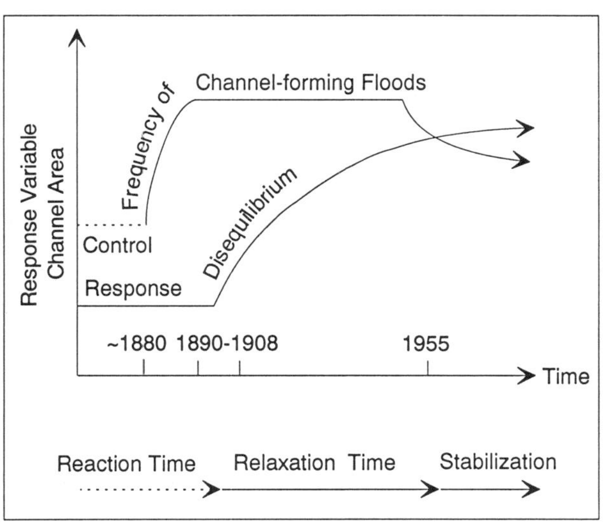

Figure 42. Analysis of channel equilibrium in terms of control and response variables.

Climate directly controls flood frequency through rainfall variations. This effect has probably been small since entrenchment, as wet-season rainfall and rainfall antecedent to the annual flood has remained constant or increased slightly. Indirectly, climate controls flood frequency through its effect on vegetation both within and out of the channel. The above-average low-intensity rainfall during the period 1957–1967, for example, might have enhanced growth of vegetation, reducing basin runoff, which in turn reduced peak flows.

Changes of grazing practices and development of water-retention structures probably shortened the time to stabilization. Generally, this resulted from improved basin conditions that reduced runoff and peak flows. The number of cattle grazing in the upper basin has decreased since entrenchment. The historic high was 36,000 cattle in 1890, which decreased gradually to 7,500 by 1964, well within grazing capacity (Rodgers, 1965, p. 68, 117, 137). During the late 1800s to the early part of the postentrenchment era, open range was the main source of grazing lands, and use of this largely unregulated resource encouraged overstocking (Wagoner, 1961; Rodgers, 1965). After passage of the Taylor Grazing Act of 1934, use of public lands was regulated and the number of cattle permitted was controlled. In addition, numerous, small water-retention structures have been built in small tributaries of the river. Although their overall effect is unknown, these stock ponds and small reservoirs were designed to reduce runoff. Construction of widely distributed stock ponds also greatly reduces the effect of grazing on riparian zones by providing uniform livestock distribution (Hendrickson and Minckley, 1984, p. 161). Thus, the density of cattle grazing along the San Pedro River has probably decreased, thereby enhancing development of riparian vegetation.

Finally, the disequilibrium period (Fig. 42) can be regarded as the complex response (Schumm, 1977, p. 13–14; 1985) of the fluvial system to the entrenchment disturbance. Although the entrenchment and widening process is erosional, sediment has accumulated on floodplains of several different ages. Thus, an erosional period has been punctuated by several depositional episodes. Equilibrium, therefore, was attained through a complex erosional and depositional process.

Implications for channel and floodplain management

Future development of the San Pedro River channel is a highly speculative topic; a number of geomorphic uncertainties permit only broad generalizations to be made (Schumm, 1985, 1991). Nonetheless, management of the resource requires general predictions regarding the stability of the channel system. Evidence indicates that the channel has or is close to a stable configuration. This new equilibrium was reached after at least 55 yr of adjustment through widening. The implication for channel and floodplain management is that the system has largely adjusted to the postentrenchment conditions. Therefore, the system will probably not change significantly, if these conditions remain within existing limits. Continued widening through bank erosion should be local and of limited extent because the banks are stabilized by vegetation. Widening is likely at the unprotected sites mainly on the concave banks of meander bends, as illustrated in Figure 5. However, even these sites may stabilize eventually as vegetation continues to spread through the system.

The effects of a catastrophic flood such as the flood of October 1983 on the Santa Cruz River (Saarinen and others, 1984) are difficult to predict. It is possible, however, that the San Pedro River channel and floodplain are reasonably stable against such a flood. Unlike the Santa Cruz River (Webb and Betancourt, 1990, p. 46), peak floods have decreased over the period of record, the result of increased vegetation and a gradually widening channel. Rainfall antecedent to the largest floods of the post-1955 era has increased over the pre-1955 era, and on two occasions has exceeded the rainfall antecedent to the flood of record in 1926. Unusually large rainfall has occurred since stabilization, but the resulting floods had relatively small peak discharge, and channel enlargement was negligible. Nevertheless, a storm such as that of October 1983 centered over the upper San Pedro River valley would be a matter of concern.

Significant aggradation and channel filling will require a shift to vertical accretion, although presently alluviation is dominated by lateral accretion. Historically, the system has the potential for aggradation after entrenchment, as the channel has been repeatedly entrenched and subsequently filled in the recent geologic past. The factors causing aggradation and the time necessary to fill the channel are essentially unknown. The most recent aggradation that ended with historic entrenchment probably lasted about 450 yr. The time necessary to fill the entrenched paleochannel is unknown. The channel could have filled soon after the shift to aggradation, because the volume of the channel is relatively small. If the present channel is in equilibrium, however, another disturbance would be required to initiate aggradation. In this case, the control variable might not be flood frequency; instead, excess sediment supply could be the controlling factor. Finally, significant erosion of the existing channel and floodplain alluvium is probably an unlikely future development. Evidence of reworking and erosion of earlier deposits such as cutbanks and other erosional features occur, but they are local and probably insignificant. The dense vegetation in the channel makes widespread reworking unlikely.

The crucial element of the stabilized channel system is the riparian forest and associated floral elements. Maintaining a healthy, reproducing riparian plant community is probably the most important management strategy; without this and other vegetation, the channel will probably change dramatically. Judging by the success of riparian vegetation at colonizing the postentrenchment channel, a viable riparian plant community has developed, which is maintained by the existing flow regimen.

The crucial elements of the undisturbed flow regimen for cottonwood reproduction are the seasonal timing and volume of water flow through the riparian community (Fenner and others, 1985). In addition, cottonwood requires an adequate supply of sediment to assure local aggradation and production of seed beds (Asplund and Gooch, 1988). Moreover, the optimal germination requirements of seedlings of the principal riparian trees of the San Pedro River are closely linked to water stress and salinity (Siegel and Brock, 1990). Thus, seed dispersion and subsequent germination are dependent on the seasonal timing of runoff, adequate runoff volume and sediment load, and unrestrained floods. Factors that reduce runoff volume, increase salinity, change runoff seasonality, or reduce sediment loads are detrimental to the riparian community.

Impoundment of sediment in reservoirs and upstream withdrawals of surface water for agriculture, mining, or domestic use will compromise the present flow regimen, degrading the recently developed riparian community. This community is also closely linked with groundwater level; a drop in this level would probably have the same effect on the riparian community as upstream impoundments and withdrawals. The effect of lowering the water table is well illustrated by the extensive degradation of the riparian environment following the entrenchment of the San Pedro River channel between 1890 and 1908. In short, extensive development and exploitation of groundwater resources will almost surely lower the water table, with predictable consequences for the riparian forest.

SUMMARY AND CONCLUSIONS

The deposits of the inner valley of the upper San Pedro River valley are subdivided into pre- and postentrenchment alluvium. The pre-entrenchment alluvium forms a terrace that occupies most of the inner valley. These deposits are late Holocene and probably correlate with the McCool Ranch member of the Escapule Ranch formation of Haynes (1987). Deposition of the McCool Ranch alluvium began about A.D. 1450; deposition probably lasted until entrenchment of the San Pedro River around turn of the century. Historic accounts of the San Pedro River suggest that cienegas, or marshy areas, were widespread before entrenchment (Hendrickson and Minckley, 1984). Evidence of the cienegas is preserved in the pre-entrenchment alluvium by one to several dark, carbonaceous beds that typically occur near the top of the unit. These beds are significant because they define the pre-entrenchment water table.

Historic accounts and photographs document the date of entrenchment and the pre-entrenchment geomorphology of the study area. Generally, the river was shallow and near the surface of the pre-entrenchment alluvium from as early as 1700 until the period about 1890–1908. The use of irrigation by Sobaipuri Indians around 1700 suggests that the channel was then unentrenched. Written accounts from 1846–1951 indicate clearly that the channel was also unentrenched at that time. Moreover, the channel was unentrenched in the study area as late as 1878.

Photographs of the downstream portion of the area in the period 1882–1890 show that the channel was not entrenched at the photograph sites. However, a photograph in the upstream portion of the area taken in 1908 indicates that the channel was recently incised. Entrenchment, therefore, occurred after about 1890 and before 1908. Thus, more than 32 km of the channel from Fairbank upstream to Hereford was probably entrenched in less than 18 years.

The immediate cause of entrenchment was a series of large floods that historic accounts suggest began in 1881. The area was also disturbed by a high-intensity earthquake in 1887 that had several documented hydrologic effects, including a fissured zone the length of the inner valley and changes in streamflow and water table. This seismic-related disruption of the channel system might have been a significant factor in preconditioning the channel for entrenchment through disruption of the water table.

The cause of the large floods is a subject of considerable debate for the San Pedro River and other streams in the Southwest, as flood-induced entrenchment was typical of the late 18th century throughout the Southwest (Cooke and Reeves, 1976; Graf, 1983). Overstocking and other human activity related to rapid settlement of the upper San Pedro River valley was coincident with the large floods and subsequent entrenchment. Many researchers have attributed entrenchment to human activity alone (see summaries in Dobyns, 1981, and Bahre, 1991); however, the effect of overgrazing is not this straightforward. Cattle have grazed the upper San Pedro River valley for at least 300 yr. Moreover, stocking levels of the early to mid-1800s during the Spanish-Mexican phase of the Arizona cattle industry (Haskett, 1936) were possibly as high as those of the entrenchment era. Thus, grazing by large numbers of animals began long before the flood-related entrenchment.

Recent work by Betancourt and Turner (1993) indicates that rainfall during the late 1800s was unusually high at Tucson. This unusual rainfall was caused by strong and frequent ENSO (El Niño Southern Oscillation) events, in a pattern that has few similarities in the 20th century. Thus, climate of the Southwest during the entrenchment era was conducive to large floods. The increased rainfall was regional and quite likely affected the upper San Pedro River valley, although weather records are unavailable for the critical period preceding entrenchment. The unusual rainfall and floods of this era occurred at the end of the Little Ice Age (Bradley, 1985), and climatic adjustment associated with the subsequent global warming is possible. In short, climate ranks closely with human activity as the principal cause of entrenchment.

The postentrenchment alluvium consists of, from oldest to youngest, terrace, floodplain, and channel of the San Pedro River. These deposits occupy the lowest topographic level of the inner valley which is 1 to 10 m below the pre-entrenchment terrace. The alluvium was deposited entirely since entrenchment, or since about 1900; however, most of the mapped alluvial units postdate 1937. A widespread, locally dense riparian forest has developed simultaneously with deposition of the postentrenchment alluvium.

The postentrenchment alluvium was deposited in an entrenched, meandering, and low-sinuosity alluvial system. The alluvium formed mainly by lateral accretion of point bars, point-bar like features, and channel bars. The deposits are fundamentally different than the pre-entrenchment alluvium. Sedimentologically, the postentrenchment alluvium is coarser grained, better sorted, and lacks carbonaceous accumulations associated with a high water table.

Alluvial sheetwash deposits and alluvial fans of mostly postentrenchment age occur on the pre-entrenchment terrace. These deposits are probably correlative with the Teviston formation of Haynes (1987). Deposition of the Teviston alluvium, however, preceded entrenchment, as the deposits are cut by the entrenched river channel. The beginning of this deposition probably records the initial disturbance that eventually led to entrenchment.

The postentrenchment alluvial deposits are successively younger across the floodplain surface, indicating that the channel has widened since initial entrenchment. The rate of channel widening was determined from the time-dependent increase of channel area. This is possible because channel area is a spatially integrated function of channel width. The area of the alluvial deposits (referred to as channel area) was measured from the mapped boundary of the entrenched channel on five sets of sequential-aerial photography taken from 1937–1986. Results indicate that channel area increased rapidly from initial entrenchment until at least 1955; since 1955 the channel area has increased only slightly. The conclusion is that the channel is largely stabilized and that equilibrium or near-equilibrium conditions exist. The relaxation time of the system, or the time to reach approximate equilibrium, was about 55 yr.

Peak-flood discharge of the San Pedro River declined substantially after 1955, approximately coincident with and probably related in part to the decline of channel enlargement. The average daily wet-season discharge of the river also declined in 1960, a result of the smaller annual floods. The historic climate of the area was analyzed to search for variations that might explain this reduced flood frequency. The analysis emphasizes wet-season (June 15–October 15) rainfall, as this season is the most significant of the year in terms of floods and runoff. Rainfall for 12 days antecedent (11 days before the flood and the day of the flood) to the annual flood was examined for the period 1916–1987 using data from eight weather stations. This analysis suggests that antecedent rainfall probably increased since 1955, whereas peak-flood and average daily discharge actually decreased.

Time series of wet-season rainfall were developed from daily weather data to search for long-term rainfall patterns. Results indicate that total wet-season rainfall has not changed significantly over the period 1897–1987. Beginning about 1951, however, the short-term pattern of wet-season rainfall probably changed. Before 1951, wet-season rainfall oscillated annually from above to below average. Since 1951, this pattern was replaced by a biennial or longer oscillation about the long-term average. The early pattern might have adversely affected vegetation by failure during a dry year of vegetation established in the preceding wet year. Grazing practices that favor adding cattle during a wet year and discourage removing them during a dry year also contributed to poor vegetation conditions.

Wet-season rainfall intensity was also analyzed for the period 1895–1987. High-intensity rainfall is defined as the seasonal frequency of daily rains ⩾95th percentile of all rainfall at a station, which is close to 2.54 cm (1.0 in.) for the eight stations. A high frequency of rains in this category is generally thought to be associated with fluvial erosion (Leopold, 1951; Cooke and Reeves, 1976; Betancourt and Turner, 1993). High-intensity rainfall, however, does not have a detectable long-term pattern in the upper San Pedro River valley, suggesting that reduced peak flows and runoff are unrelated to fewer high-intensity rainfall events.

The relation between the duration of large runoff and the duration of high-intensity rainfall has probably changed since 1960. This rainfall category is significantly and moderately well correlated ($r^2 = 0.43$) with the seasonal duration of large runoff, defined as runoff ⩾95th percentile of all wet-season daily discharge. On average, 1 to 2 days of high-intensity rainfall annually produced about 7 to 8 days of large runoff before 1960. After 1960, however, 1 to 2 days of high-intensity rainfall produced only about 3 to 4 days of large runoff. The duration of high-intensity rainfall evidently did not change, but runoff duration shortened. This shortened duration probably resulted partly from increased channel sinuosity and vegetation in the channel, which increased transmission losses.

Low-intensity rainfall is defined as the seasonal frequency of rainfall <80th percentile of all rainfall at a station; this is close to 1.27 cm (0.5 in.) for the eight stations. A low frequency of rains in this category is thought to adversely effect plant growth, whereas a relatively large number of days with such rainfall is thought to promote growth. Low-intensity rainfall was below to only slightly above the long-term average frequency during 1897–1928, the period of entrenchment and rapid channel widening. In contrast, this rainfall category was consistently above the long-term average frequency during 1957–1967, the beginning of the period of reduced channel enlargement. The reduced peak flows of the post-1955 era might have resulted partly from an improved vegetation cover in the basin related to high frequency of low-intensity rainfall.

In short, the channel of the San Pedro River has probably stabilized after at least 55 yr of instability. Stabilization occurred in the absence of any detectable long-term shift of rainfall patterns. However, an increased frequency of wet season low-intensity rainfall in the late 1950s to 1960s, and a shift to extended periods of normal to above normal rainfall might have hastened stabilization. Stabilization was also hastened by improved land use practices and other conservation measures. Stocking levels, for example, declined substantially from 1900 until at least 1964 (Rodgers, 1965, p. 137). Moreover, stock ponds and other small water-retention structures have been constructed on small tributaries of the San Pedro River, and pondage in these structures would also reduce peak flows.

ACKNOWLEDGMENTS

A number of persons have contributed their time, information, and ideas. Julio L. Betancourt, Raymond M. Turner, and Robert H. Webb supplied unpublished photographs of the channel from their files and a portion of the 1937 aerial photography. A field trip and numerous conversations with Betancourt regarding the Santa Cruz and San Pedro Rivers improved my thinking about climate and stream entrenchment in southeast Arizona. Several days in the field with Edmund D. Andrews helped sharpen my thoughts concerning the San Pedro River. Tina A. Bauer compiled and measured channel area and assisted with field work in the late winter of 1990. Ben Lomeli also helped with several days of field work in 1990 and has been a constant source of useful information regarding the river. The critical reviews and comments of Edmund D. Andrews, Victor R. Baker, William L. Graf, William L. Jackson, and William C. Johnson contributed to the report. This research was funded through a cooperative agreement between the U.S. Geological Survey and the U.S. Bureau of Land Management.

REFERENCES CITED

Andrade, E. R., and Sellers, W. D., 1988, El Niño and its effect on precipitation in Arizona and western New Mexico: Journal of Climatology, v. 8, p. 403–410.

Asplund, K. K., and Gooch, M. T., 1988, Geomorphology and the distributional ecology of Fremont cottonwood (*Populus fremontii*) in a desert riparian canyon: Desert Plants, v. 9, p. 17–26.

Bahre, C. J., 1991, A legacy of change: Tucson, University of Arizona Press, 231 p.

Bahre, C. J., and Bradbury, D. E., 1978, Vegetation change along the Arizona-Sonora boundary: Annals of the Association of American Geographers, v. 68, p. 145–165.

Bahre, C. J., and Hutchinson, C. F., 1985, The impact of historic fuelwood cutting on the semidesert woodlands of southeastern Arizona: Journal of Forest History, v. 29, p. 176–186.

Bartlett, J. R., 1854, Narrative of explorations and experiences in Texas, New Mexico, California, Sonora, and Chihuahua, Vol. I: New York, D. Appleton and Company.

Betancourt, J. L., 1986, Historic flow conditions and channel changes on the San Pedro River, southern Arizona: Unpublished report available from author, 29 p.

Betancourt, J. L., 1990, Tucson's Santa Cruz River and the arroyo legacy [unpublished Ph.D. dissertation]: Tucson, University of Arizona, 237 p.

Betancourt, J. L., and Turner, R. M., 1993, Tucson's Santa Cruz River and the arroyo legacy: Tucson, University of Arizona Press (in press).

Bolton, H. E., 1936, Rim of Christendom: A biography of Eusebio Francisco Kino: Tucson, University of Arizona Press, 644 p.

Bradley, R. S., 1985, Quaternary paleoclimatology: Boston, Allen and Unwin, 472 p.

Burkham, D. E., 1976, Effects of changes in an alluvial channel on the timing, magnitude, and transformation of flood waves, southeastern Arizona: U.S. Geological Survey Professional Paper 655-K, p. K1–K25.

Burkham, D. E., 1981, Uncertainties resulting from changes in river form: American Society of Civil Engineers, Journal of the Hydraulics Division, v. 107, HY5, p. 593–610.

Carleton, A. M., 1986, Synoptic-dynamic character of "bursts" and "breaks" in the south-west U.S. summer precipitation singularity: Journal of Climatology, v. 6, p. 605–623.

Christiansen, L. D., 1983, The Mormon battalion in Cochise County and adjacent areas: Cochise Quarterly, v. 13, 47 p.

Christiansen, L. D., 1988, The extinction of wild cattle in southern Arizona: Journal of Arizona History, v. 29, p. 89–100.

Conkling, R. P., and Margaret, B., 1947, The Butterfield overland mail: Glendale, California, Arthur H. Clark Co., p. 147.

Cooke, P. St. G., 1938, Cookes journal of the march of the Mormon battalion 1846–1847, eds., Bieber, R. P., and Averam, B. B., 1938: Glendale, California Arthur H. Clark Co., 100 p.

Cooke, R. U., and Reeves, R. W., 1976, Arroyos and environmental change in the American south-west: Oxford, England, Clarendon Press, 213 p.

Cox, J. R., and Jordan, G. L., 1983, Density and production of seeded range grasses in southeastern Arizona (1970–1982): Journal of Range Movement, v. 36, p. 649–652.

Davey, J. R., 1980, Energy flow dynamics of a desert grassland ecosystem [unpublished Ph.D. dissertation]: Northern Arizona University, 147 p.

Dobyns, H. F., 1981, From fire to flood: Historic human destruction of Sonoran Desert riverine oases: Socorro, New Mexico, Ballena Press, 222 p.

DuBois, S. M., and Smith, A. W., 1980, The 1887 earthquake in San Bernardino valley, Sonora: Historic accounts and intensity patterns in Arizona: Tucson, Arizona Bureau of Geology and Mineral Technology Special Paper 3, 112 p.

Elliot, J. G., 1979, Evolution of large arroyos—The Rio Puerco of New Mexico [unpublished M.S. thesis]: Fort Collins, Colorado State University, 106 p.

Fenner, P., Ward, W. B., and Patton, D. R., 1985, Effects of regulated water flows on regeneration of fremont cottonwood: Journal of Range Management, v. 32, p. 135–138.

Frasier, G. W., and Woolhiser, D. A., 1990, Precipitation frequency effects on seedling survival, *in* Unger and others, eds., Challenges in dryland agriculture—A global perspective: Amarillo, Texas, Bushland, Proceedings International Conference on Dryland Farming, p. 310–312.

Fulton, R. W., 1966, Milleville-Charleston, Cochise County, 1878–1889: Journal of Arizona History, v. 7, p. 9–22.

Gehlbach, F. R., 1981, Mountain islands and desert seas: College Station, Texas A & M University Press.

Gifford, G. F., and Hawkins, R. H., 1978, Hydrologic impact of grazing on infiltration: A critical review: Water Resources Research, v. 14, p. 305–313.

Graf, W. L., 1977, The rate law in fluvial geomorphology: American Journal of Science, v. 277, p. 178–191.

Graf, W. L., 1983, The arroyo problem: Paleohydrology and paleohydraulics in the short term, *in* Gregory, K. G., ed., Background to paleohydrology: New York, John Wiley, p. 279–302.

Graf, W. L., 1988, Fluvial processes in dryland rivers: New York, Springer-Verlag, 346 p.

Graham, J. L., 1976, Ghost towns of Cochise County: Cochise Quarterly, v. 6, p. 12–27.

Gray, R. S., 1965, Late Cenozoic sediments in the San Pedro Valley near St. David, Arizona [unpublished Ph.D. Dissertation]: Tucson, University of Arizona, 198 p.

Green, C. R., and Sellers, W. D., 1964, Arizona climate: Tucson, University of Arizona Press, 503 p.

Harvey, M. D., Watson, C. C., and Schumm, S. A., 1985, Gully erosion: Bureau of Land Management Technical Note 366, 181 p.

Haskett, B., 1935, Early history of the cattle industry in Arizona: Arizona Historical Review, v. 6, p. 3–42.

Hastings, J. R., and Turner, R. M., 1965, The changing mile: Tucson, University of Arizona Press, 317 p.

Haynes, C. V., Jr., 1987, Curry Draw, Cochise County, Arizona: A late Quaternary stratigraphic record of Pleistocene extinction and paleo-Indian activities, *in* Davis, G. H., and VandenHolder, E. M., eds., Geologic diversity of Arizona and its margins: Excursions to choice areas: Tucson, Arizona Bureau of Geology and Mineral Technology Special Paper 5, p. 23–28.

Hendrickson, D. A., and Minckley, W. L., 1984, Cienegas—Vanishing climax communities of the American Southwest: Desert Plants, v. 6, p. 131–175.

Hereford, R., 1984, Climate and ephemeral-stream processes: Twentieth century geomorphology and alluvial stratigraphy of the Little Colorado River, Arizona: Geological Society of America Bulletin, v. 95, p. 654–668.

Hereford, R., 1986, Modern alluvial history of the Paria River drainage basin, southern Utah: Quaternary Research, v. 25, p. 293–311.

Hereford, R., 1989, Variation of warm-season rainfall frequency in the southern Colorado Plateau and its effect on runoff and alluvial channel morphology: U.S. Geological Survey Open-File Report 89-330, 16 p.

Hereford, R., and Webb, R. H., 1992, Historic variation of warm-season rainfall, southern Colorado Plateau, southwestern U.S.A.: Climatic Change, v. 22, p. 239–256.

Hinton, R. J., 1978, The hand-book to Arizona: Tucson, Arizona Silhouettes, p. 231–235.

Hunt, C. E., 1988, Down by the river; The impact of Federal water projects and policies on biological diversity: Island Press, p. 81–86.

Ivey, J. E., Rhodes, D. L., and Sanchez, J. P., 1991, The Coronado Expedition of 1540–1542, A special history report prepared for the Coronado Trail study: Denver, Colorado, U.S. Department of Interior, National Park Service, 154 p.

Jackson, W. L., and 7 others, 1987, Assessment of water conditions and management opportunities in support of riparian values: BLM San Pedro River properties, Arizona, project completion report: Denver, Colorado, U.S. Department of Interior, Bureau of Land Management Service Center, 180 p.

Kessell, J. L., 1966, The puzzling presido San Phelipe de Guevavi, alias Terrenate: New Mexico Historical Review, v. 4, p. 20–46.

Knighton, D., 1984, Fluvial forms and processes: London, England, Edward Arnold, 218 p.

Lane, L. L., 1990, Transmission losses, flood peaks, and groundwater recharge, *in* French, R. H., ed., Hydraulics/hydrology of arid lands: New York, Hydraulics Division of the American Society of Civil Engineers, p. 343–348.

Leopold, L. B., 1951, Rainfall frequency: An aspect of climatic variation: EOS, Transactions of the American Geophysical Union, v. 32, p. 347–357.

Mattison, R. H., 1946, Early Spanish and Mexican settlements in Arizona: New Mexico Historical Review, v. 21, p. 272–327.

McClintock, J. H., 1921, Mormon settlement in Arizona: Phoenix, Arizona, Manufacturers Stationers.

Meyer, D. F., 1989, The significance of sediment transport in arroyo development: U.S. Geological Survey Water-Supply Paper 2349, 50 p.

Munson, R. W., 1976, Don Ignacio and Doña Eulalia Elias and the history of their hacienda on the Babocomari: Camp Wallen: Cochise Quarterly, v. 6, p. 3–11.

Myrick, D. F., 1975, Railroads of Arizona, Vol. I, The southern roads: Berkeley, California, Howell North, 477 p.

NOAA, 1986, Surface land daily cooperative summary of the day TD-3200: Asheville, North Carolina, National Climate Data Center, 23 p.

Osborn, H. B., and Lane, L. J., 1984, Climatic change and streamflow in the Southwest, *in* Replogle, J. A., Renard, K. G., ed., Water today and tomorrow: Flagstaff, Arizona, Proceedings, American Society of Civil Engineers Specialty Conference, Irrigation and Drainage Division, p. 362–371.

Rodgers, W. M., 1965, Historical land occupance of the upper San Pedro River valley since 1870 [unpublished M.S. thesis]: University of Arizona, Tucson, 167 p.

Roeske, R. H., Garrett, J. M., and Eychaner, J. H., 1989, Floods of October 1983 in southeastern Arizona: U.S. Geological Survey Water-Resources Investigations Report 85-4225-C, 77 p.

Ropelewski, C. F., and Halpert, M. S., 1986, North American precipitation and temperature patterns associated with the El Niño Southern Oscillation (ENSO): Monthly Weather Reviews, v. 114, p. 2352–2362.

Rushforth, S. R., and Brotherson, J. D., 1982, Cryptogamic soil crusts in the deserts of North America: American Biology Teacher, v. 44, p. 472–475.

Saarinen, T. F., Baker, V. R., Durrenberger, R., and Maddock, T., Jr., 1984, The Tucson, Arizona flood of October 1983: Washington, D.C., National

Academy Press, 112 p.
Schumm, S. A., 1977, The fluvial system: New York, John Wiley & Sons, 338 p.
Schumm, S. A., 1985, Explanation and extrapolation in geomorphology: Seven reasons for geologic uncertainty: Transactions Japanese Geomorphological Union, v. 6, p. 1–18.
Schumm, S. A., 1991, To interpret the Earth: Ten ways to be wrong: New York, New York, Cambridge University Press, 133 p.
Sellers, W. D., and Hill, R. H., 1974, Arizona climate: Tucson, University of Arizona Press, 616 p.
Siegel, R. S., and Brock, J. H., 1990, Germination requirements of key southwestern woody riparian species: Desert Plants, v. 10, p. 3–8.
Thornber, J. J., 1910, The grazing ranges of Arizona: Arizona Agricultural Experiment Station Bulletin No. 65, p. 335–340.
Tiller, K. S., 1982, Charleston townsite revisited: Journal of Arizona History, v. 23, p. 241–248.
Toumey, J. W., 1891, Overstocking the range: Arizona Agricultural Experiment Station Bulletin No. 2, p. 7–8.
Trischka, C., 1971, A history of Cochise County, Arizona: Cochise Quarterly, v. 1, p. 3–21.
Wagoner, J. J., 1962, Overstocking of the ranges in southern Arizona during the 1870's and 1880's: Arizoniana, v. 2, p. 23–27.
Waters, M. R., 1985, Late Quaternary stratigraphy of Whitewater Draw, Arizona: Implications for regional correlation of fluvial deposits in the American Southwest: Geology, v. 13, p. 705–708.
Waters, M. R., 1988, Holocene alluvial geology and geoarcheology of the San Xavier reach of the Santa Cruz River, Arizona: Geological Society of America Bulletin, v. 100, p. 479–491.
Webb, R. H., 1985, Late Holocene flooding on the Escalante River, south-central Utah: [unpublished Ph.D. dissertation]: Tucson, University of Arizona, 204 p.
Webb, R. H., and Betancourt, J. L., 1990, Climatic variability and flood frequency of the Santa Cruz River, Pima County, Arizona: U.S. Geological Survey Open-File Report 90-553, 69 p.

Manuscript Accepted by the Society October 19, 1992

Printed in U.S.A.

Typeset by WESType Publishing Services, Inc., Boulder, Colorado
Printed in U.S.A. by Malloy Lithographing, Inc., Ann Arbor, Michigan